Fluorine Chemistry
for Organic Chemists

Fluorine Chemistry
for Organic Chemists
Problems and Solutions

MILOŠ HUDLICKÝ

New York   Oxford
OXFORD UNIVERSITY PRESS
2000

Oxford University Press

Oxford   New York
Athens   Auckland   Bangkok   Bogotá   Buenos Aires   Calcutta
Cape Town   Chennai   Dar es Salaam   Delhi   Florence   Hong Kong   Istanbul
Karachi   Kuala Lumpur   Madrid   Melbourne   Mexico City   Mumbai
Nairobi   Paris   São Paulo   Singapore   Taipei   Tokyo   Toronto   Warsaw

and associated companies in
Berlin   Ibadan

Copyright © 2000 by Miloš Hudlický

Published by Oxford University Press, Inc.
198 Madison Avenue, New York, New York 10016

Library of Congress Cataloging-in-Publication Data
Hudlický, Miloš, 1919–
      Fluorine chemistry for organic chemists : problems and solutions / Miloš Hudlický.
         p.   cm.
      Includes bibliographical references and index.
      ISBN 0-19-513156-8
      1. Organofluorine compounds. I. Title.
QD305.H15H826   2000
547′.02—dc2   99-24931

1 3 5 7 9 8 6 4 2

Printed in the United States of America
on acid–free paper

To my dearest wife Alena
and to our children, Tomáš and Eva.

Preface

The chemistry of fluorine and its compounds has long lagged behind the chemistry of the other three halogens. While hydrochloric acid was well known in antiquity, hydrogen fluoride was first described by Scheele only in 1771, and fluorine was first prepared by H. Moissan in 1886. Organic fluorine compounds were studied systematically by F. Swarts only at the end of the last century, and the first industrial application of fluorine compounds came from the discovery of fluorinated refrigerants by A. L. Henne and T. Midgley around 1930.

The initial delay in the development of the chemistry of fluorine and its compounds compared with the chemistry of other halogens was soon erased by very intensive work in fluorine chemistry. Hundreds of thousands of fluorine compounds have been prepared over the last century, and the number of fluorine chemistry monographs exceeds by many times that of other halogens.

Some of the fluorinated compounds possess unique properties. Although most reactions of fluorine compounds resemble those of chlorine, bromine, and iodine, in many cases fluorinated compounds show peculiar dissimilarities. Such dissimilarities are the subject of this volume.

This book consists of two parts—Problems and Solutions. Each part is further subdivided according to the types of reactions. Entries in the Contents feature two page references, separated by a slash. The first page number refers to the page on which the problem is posed; the second page number refers to the page on which the solution to the problem is described.

Acknowledgments   The author thanks his colleagues Drs. Mike Calter and Paul Deck for reading and commenting on the manuscript, and Ms. Angie Miller and Ms. April Miller for preparing the camera-ready copy.

Contents

Part I. Problems

1 Warmup for Inorganic Chemists   3 / 41

1. Safety of Hydrogen Fluoride Cylinders   3 / 41
2. Bond Dissociation Energy of the Molecule of Fluorine   3 / 42

2 Introduction of Fluorine   4 / 42

3. Addition of Fluorine to Diphenylethylenes   4 / 42
4. Rearrangement in the Replacement of Hydroxyl by Fluorine in Saccharides   4 / 44
5. Reaction of β-Dicarbonyl Compounds with Diethylaminosulfur Trifluoride   5 / 45

3 Reductions   6 / 46

6. Reduction of Unsaturated Fluorinated Compounds   6 / 46
7. Reduction of 3-Chlorotetrafluoropyridine   6 / 47
8. Reaction of Isopropyl Alcohol with 2-Fluorocyclohexanone   6 / 48

4 Oxidations   7 / 49

9. Oxidation of Chlorotrifluoroethylene with Oxygen   7 / 49
10. Reaction of Perfluoro-o-phenylenediamine with Lead Tetraacetate   7 / 49
11. Reaction of Fluorinated Aromatic Diamines with Nitrous Acid   8 / 50

5 Preparation of Halogen Derivatives   8 / 51

12. Addition of Iodine Fluoride to Chlorotrifluoroethylene   8 / 51
13. Addition of Hydrogen Bromide to Chlorotrifluoroethylene   8 / 52
14. Addition of Hydrogen Chloride to 3,3,3-Trifluoropropene   9 / 53
15. Addition of Hydrogen Halides to Perfluoropropene   9 / 53
16. Reaction of Perfluoroalkenes with Metal Fluorides   9 / 54
17. Chlorination of o-Fluorotoluene   9 / 55
18. Chlorination of 3,4-Difluoronitrobenzene   10 / 56
19. Reaction of Fluoropentanitrobenzene with Hydrogen Chloride   10 / 56
20. Reaction of 1,1-Difluoroethylene with Trifluoroacetyl Hypochlorite   10 / 57
21. Reaction of Pentafluorophenol with tert-Butyl Hypobromite   11 / 57
22. Reaction of Perfluorocycloalkenes with Aluminum Halides   11 / 58
23. Reaction of Perfluoro-2-butyltetrahydrofuran with Aluminum Chloride   11 / 58
24. Reaction of Perfluorotetrahydropyran with Aluminum Chloride   12 / 58

6 Nitration 12 / 59

25. Nitration of *o*-Fluorotoluene   12 / 59

7 Reactions of Sulfur Trioxide 12 / 60

26. Reaction of Sulfur Trioxide with Enol Ethers   12 / 60
27. Reaction of Sulfur Trioxide with Perfluoro-*tert*-butylacetylene   13 / 60
28. Reaction of Sulfur Trioxide with Perfluoroisobutylene   13 / 61
29. Reaction of Sulfur Trioxide with Perfluoropropylene Oxide   13 / 62
30. Reaction of Sulfur Trioxide with Perfluorotoluene   13 / 62

8 Acid-Catalyzed Additions and Substitutions 14 / 62

31. Reaction of Benzene with 1-Chloro-2-fluoropropane   14 / 62
32. Reaction of Benzene with 2-Chloro-1,1,1-trifluoropropane   14 / 63
33. Reaction of Benzene with 3,3,3-Trifluoropropene   14 / 64
34. Reaction of 1-Phenylperfluoropropene with Aluminum Chloride   14 / 64

9 Hydrolyses 15 / 65

35. Hydrolysis of Benzyl Halides   15 / 65
36. Hydrolysis of ω-Fluorocarboxylic Acids and Their Nitriles   15 / 65
37. Hydrolysis of Perfluorocycloalkenes   16 / 66
38. Reaction of 3,5-Dichlorotrifluoropyridine with Potassium Hydroxide   16 / 67
39. Reaction of 2,5-bis(Trifluoromethyl)aniline with Sodium Hydroxide   16 / 68
40. Reaction of Heptafluorobutyraldehyde Hydrate with Potassium Hydroxide   17 / 69
41. Reaction of Pentafluoroethyl Iodide with Potassium Hydroxide and Water   17 / 69
42. Hydrolysis of Unsaturated Fluorinated Compounds   17 / 70
43. Hydrolysis of Geminal Difluorides   17 / 71
44. Hydrolysis of Difluoromethylene Group in Enol Ethers   18 / 71
45. Reaction of Alkalis with Fluorinated Cyclopropanes   18 / 72
46. Product of Treatment of *p*-Trifluoromethylphenol with Potassium Hydroxide   18 / 72

10 Alkylations 19 / 73

47. Reaction of 1-Chloro-3,3,4,4-tetrafluorocyclobutene with Potassium Hydroxide and Ethanol   19 / 73
48. Reaction of 1-Bromo-2-chlorotetrafluorocyclobutene with Potassium Hydroxide   19 / 74
49. Reaction between 1-Chloro-2-fluoroethane and Ethyl Acetoacetate   19 / 75

50. Reaction of Pentafluorophenol with Chlorodifluoromethane and Sodium Hydroxide   20 / 75

51. Reaction of 1,4-Dibromohexafluoro-2-butene with Sodium Ethoxide   20 / 76

52. Reaction of Perfluoro-3,4-dimethyl-3-hexene with Methanol   20 / 76

53. Reaction of Chlorotrifluoroethylene with Sodium Ethoxide   20 / 77

54. Reaction of Perfluoro-γ-butyrolactone with Perfluoropropylene Oxide   21 / 77

55. Reaction of Alkali Thiophenoxides with Polyfluorohalomethanes   21 / 78

56. Reaction of Perfluorodecalin with Sodium Thiophenoxide   21 / 79

57. Reaction of Perfluorocyclobutene with Hydrazine   22 / 80

58. Reaction of 1,2,3,5-tetrakis(Trifluoromethyl)benzene with Ammonia   22 / 81

59. Reaction of Perfluoroisobutylene with Dimethyl Malonate   22 / 82

60. Reaction of Perfluorobenzotrichloride and Chlorodifluoromethane   22 / 83

61. Reaction of Enamines with Trifluoromethanesulfenyl Chloride   23 / 83

62. Reaction of 1,1-Dichlorodifluoroethylene with Methanol   23 / 84

63. Reaction of Chlorotrifluoroethylene with Aniline   23 / 85

64. Reaction of Chlorotrifluoroethylene with Dimethylamine   23 / 85

65. Reaction of Alcohols with Perfluoroalkenes   24 / 86

66. Reaction of Fluoroalkenes with Azides   24 / 87

11 Arylations   24 / 88

67. Reaction of 1,2-Difluorotetrachlorobenzene with Sodium Methoxide   24 / 88

68. Reaction of 2,4-Dinitrohalobenzenes with Sodium Methoxide   25 / 89

69. Reaction of Fluoropentachlorobenzene with Potassium p-Methoxyphenoxide   25 / 90

70. Reaction of Perfluoronaphthalene with Sodium Alkoxides or Aryloxides   25 / 90

71. Reaction of Perfluoropyridine with Methanol   26 / 91

72. Arylation at Sulfur   26 / 91

12 Acylations   26 / 92

73. Acylations with Mixed Anhydrides of Trifluoroacetic Acid   26 / 92

13 Aldol-Type Condensations   27 / 93

74. Reaction of 1,1,1-Trifluoroacetone, Formaldehyde, and Piperidine   27 / 93

75. Reaction of α,α,α -Trifluoroacetophenone with
Tributylphosphine   27 / 93

76. Reaction of Dibromodifluoromethane and
tris(Dimethylamino)phosphine with Fluorinated Ketones   27 / 94

77. Reactions of Phosphorus Ylides with Fluorinated Nitriles   28 / 95

78. Reaction of tris(Trifluoromethyl)methane with
Acrylonitrile   28 / 95

14 Organometallic Syntheses   28 / 95

79. Reaction of Ethyl Chlorofluoroacetate with Grignard
Reagents   28 / 95

80. Reaction of 3-Chloropentafluoropropene with Phenylmagnesium
Bromide   29 / 96

81. Reaction of 1,2-Dichlorohexafluorocyclopentene with
Ethylmagnesium Bromide   29 / 97

82. Reaction of 1,2-Dichloro-1-fluoroethylene with Butyllithium and
Acetone   29 / 98

83. Reaction of Trifluoroethylene with Butyllithium and
Acetone   30 / 98

84. Reaction of 1-Chloro-1,2-difluoroethylene with Butyllithium and
Carbon Dioxide   30 / 98

85. Reaction of Aldehydes with 1,1,1-Trichlorotrifluoroethane and
Zinc in the Presence of Acetic Anhydride   30 / 99

86. Reaction of Aldehydes with 1,1,1-Trichlotrifluororoethane and
Zinc in the Presence of Aluminum Chloride   30 / 99

87. Reaction of Dibromodifluoromethane and
4-Chloro-3-nitrobenzotrifluoride with Copper   31 / 100

15 Additions   31 / 101

88. Reaction of 2-Aminoethanol with Ethyl
4,4,4-Trifluoro-3-trifluoromethyl-2-butenoate   31 / 101

89. Reaction of 1,1-Dichlorodifluoroethylene with
1,3-Butadiene   31 / 102

90. Reaction of 1,2-Dichlorodifluoroethylene with Perfluoropropylene
Oxide   32 / 103

91. Reaction between Perfluorovinylsulfur Pentafluoride and
1,3-Butadiene   32 / 103

92. Reaction between Trifluoroethylene and 1,3-Butadiene   32 / 103

93. Reaction between Hexafluoroacetone Azine and
Acetylene   32 / 104

94. Reaction of tris(*tert*-Butyl)azete and Trifluoroacetonitrile   33 / 104

95. Dimerization of Perfluoro-1,3-butadiene   33 / 105

16 Eliminations   33 / 105

96. Reaction of Alkalis with Fluorinated Cyclopropanes   33 / 105

97. Reaction of Diethyl *threo*-2-Bromo-3-fluorosuccinate with
Potassium Acetate   34 / 106

98. Reaction of 1-Bromo-2-fluorocyclohexanes with Bases   34 / 107

17 Rearrangements   35 / 108

99. Reaction of Fluorohaloethanes with Aluminum Chloride   35 / 108
100. Treatment of 3-Chloropentafluoropropene with Antimony Halides   35 / 109
101. Reaction of Trifluoronitrosomethane with Ammonia   35 / 109
102. The Hofmann Degradation of Perfluorobutyramide   35 / 109
103. Irradiation of hexakis(Trifluoromethyl)benzene   36 / 110
104. Reaction of 2-Bromoperfluoronaphthalene with Antimony Pentafluoride   36 / 110
105. Claisen Rearrangement of Fluoroaromatic Acetylenic Ethers   37 / 111

Part II. Solutions

REFERENCES	113
AUTHOR INDEX	118
SUBJECT INDEX	121

Part I
Problems

1

Warmup for Inorganic Chemists

SURPRISE 1

Safety of Hydrogen Fluoride Cylinders

It is well known that anhydrous nonoxidizing inorganic acids such as hydrogen chloride and hydrogen bromide do not dissolve metals. These two compounds are stored in steel cylinders without any appreciable corrosion of the metal. The same was true of anhydrous hydrogen fluoride that was stored and delivered in steel cylinders for the past seven decades.

However, recently several instances have been noted of high pressure inside steel cylinders containing anhydrous hydrogen fluoride, pressure considerably higher than that corresponding to the vapor pressure of hydrogen fluoride at room temperature (1 atm at 20°C). In one case, the pressure inside a cylinder of hydrogen fluoride stored for 20–24 years was 154 atm (2200 psig); in another case it was 168 atm (2400 psig) after 14 years of storage.

SURPRISE 2

Bond Dissociation Energy of the Molecule of Fluorine

The bond dissociation energies of chlorine, bromine, and iodine have been known for a long time. In Linus Pauling's classical book *The Nature of the Chemical Bond* (1948), the bond dissociation energies of halogens are listed as 57.8, 46.1, and 36.2 kcal/mol (242, 193, and 152 kJ) for chlorine, bromine, and iodine, respectively. What is the bond dissociation energy of fluorine?

SURPRISE 3

Addition of Fluorine to Diphenylethylenes

When *cis*-stilbene is treated with elemental fluorine, a mixture of 79% *meso*-1,2-difluoro-1,2-diphenylethane and 16% (±)-1,2-difluoro-1,2-diphenylethane is obtained.

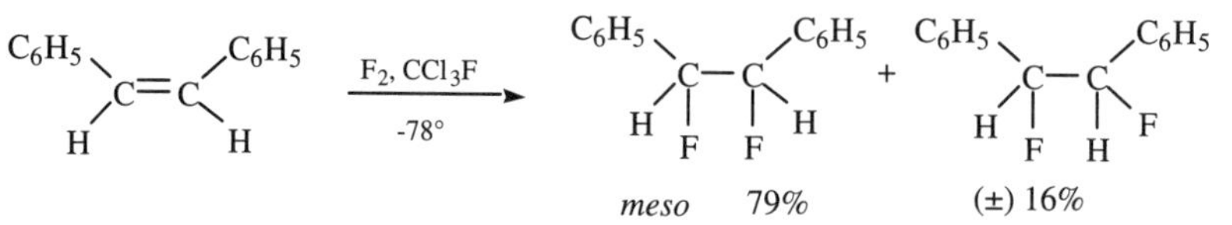

If 1,1-diphenylethylene is treated with elemental fluorine under the same reaction conditions, $C_{14}H_{12}F_2$ (compound **A**) is obtained in a 14% yield (accompanied by 78% of compound **B**). If the same compound is treated with hydrogen fluoride and lead tetrafluoride or with phenyl iodide difluoride, $C_{14}H_{12}F_2$ (compound **C**) is produced in 27% and 47% yields, respectively. Compounds **A** and **C** are different from each other, and different from both *meso*- and (±)-1,2-difluoro-1,2-diphenylethane. What are the compounds **A**, **B**, and **C**, and how are they formed?

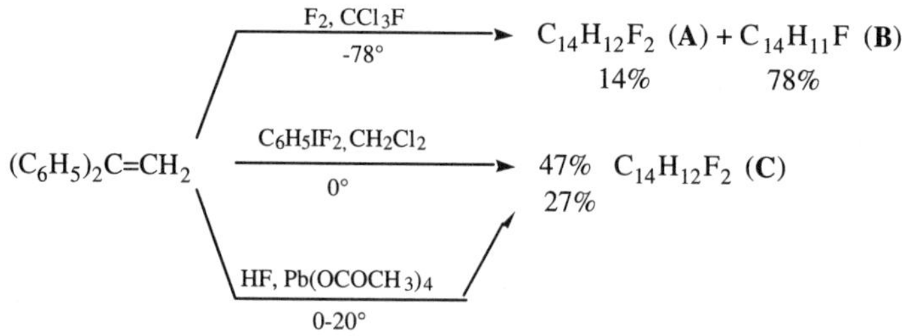

SURPRISE 4

Rearrangement in the Replacement of Hydroxyl by Fluorine in Saccharides

When benzyl 3-azido-3-deoxy-4,6-*O*-benzylidene-α-D-altropyranoside (compound I) is treated with diethylaminosulfur trifluoride (DAST), three products are obtained: II, III, and IV. Compound II results from replacement of the 2-hydroxy group by fluorine with retention of configuration; compound III is a result of rearrangement of the azido group and replacement of the 2-hydroxyl by fluorine; and compound IV involves rearrangement of the 1-benzyloxy group and replacement of the 2-hydroxyl by fluorine. How can the results of this reaction be accounted for?

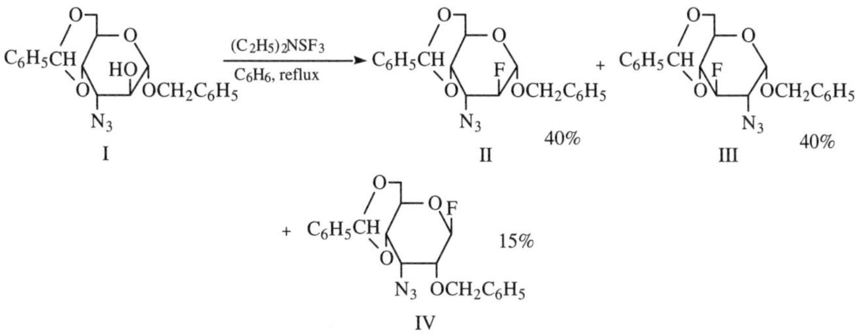

SURPRISE 5

Reaction of β-Dicarbonyl Compounds with Diethylaminosulfur Trifluoride

The usual reaction of diethylaminosulfur trifluoride (DAST) with a carbonyl group is the replacement of carbonyl oxygen by two atoms of fluorine, thus generating geminal difluorides.

However, when ethyl acetoacetate was treated with diethylaminosulfur trifluoride, an entirely unexpected difluoro compound, $C_6H_8F_2O_2$, was obtained. The compound was an ester, and decolorized a solution of bromine in carbon tetrachloride. What is the structure of the product?

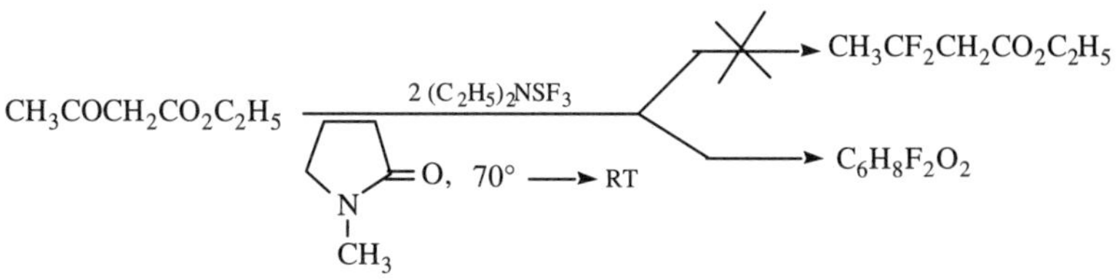

3 Reductions

SURPRISE 6

Reduction of Unsaturated Fluorinated Compounds

Catalytic hydrogenation of *trans*-α-fluorostilbene and of difluoromaleic acid does not give products expected from the addition of hydrogen across the double bonds. What are the products of catalytic hydrogenation of compounds **D**, and **E**, and **F**?

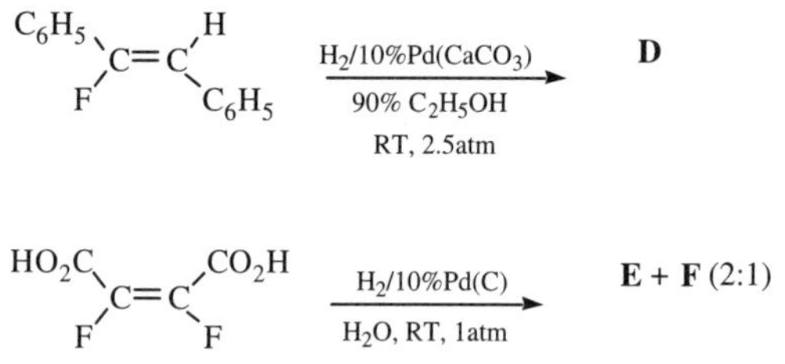

$$C_6H_5 \diagdown C = C \diagup H \atop F \diagup \diagdown C_6H_5 \quad \xrightarrow[\substack{90\% \ C_2H_5OH \\ RT, \ 2.5atm}]{H_2/10\%Pd(CaCO_3)} \quad \mathbf{D}$$

$$HO_2C \diagdown C = C \diagup CO_2H \atop F \diagup \diagdown F \quad \xrightarrow[\substack{H_2O, \ RT, \ 1atm}]{H_2/10\%Pd(C)} \quad \mathbf{E} + \mathbf{F} \ (2:1)$$

SURPRISE 7

Reduction of 3-Chlorotetrafluoropyridine

Catalytic hydrogenation of 3-chlorotetrafluoropyridine gives compound **G**, and reduction of the same starting material with complex hydrides gives compound **H**. What are compounds **G** and **H**?

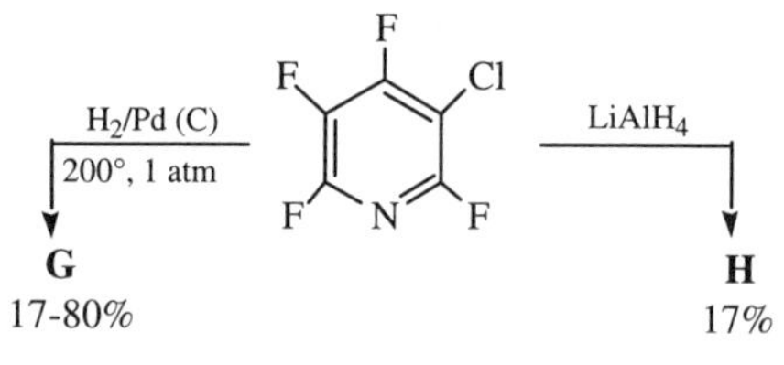

$$\xleftarrow[\substack{200°, \ 1 \ atm}]{H_2/Pd \ (C)} \qquad \text{(structure)} \qquad \xrightarrow{LiAlH_4}$$

G   **H**
17-80%   17%

SURPRISE 8

Reaction of Isopropyl Alcohol with 2-Fluorocyclohexanone

What is the product of a reaction of 2-fluorocyclohexanone with isopropyl alcohol under ultraviolet irradiation (compound **I**), and what is the product of the reaction of the same alcohol with methyl 3,3-difluoro-2,2,3-trichloropropionate under the same conditions (compound **J**)?

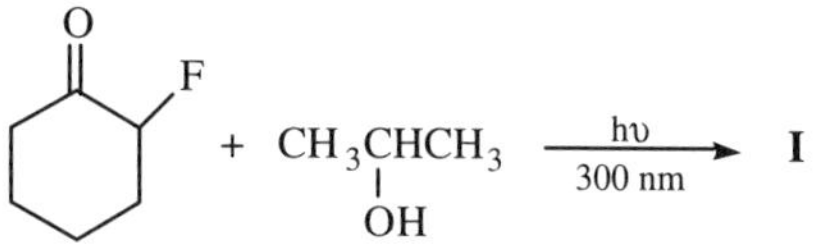

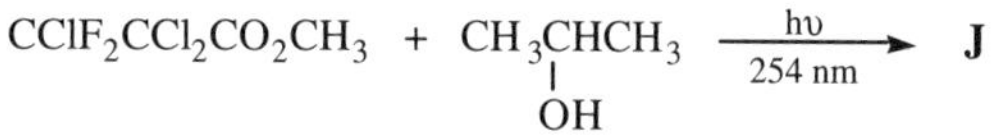

$$\text{(cyclohexanone with F at 2-position)} + \underset{\underset{OH}{|}}{CH_3CHCH_3} \xrightarrow[300\ nm]{h\upsilon} \mathbf{I}$$

$$CClF_2CCl_2CO_2CH_3 + \underset{\underset{OH}{|}}{CH_3CHCH_3} \xrightarrow[254\ nm]{h\upsilon} \mathbf{J}$$

4   Oxidations

SURPRISE 9

Oxidation of Chlorotrifluoroethylene with Oxygen

Chlorotrifluoroethylene reacts with oxygen to form compound **K**, which rearranges to compound **L**. The same compound **L** is also generated when a mixture of chlorotrifluoroethylene and oxygen is irradiated by ultraviolet light. What are compounds **K** and **L**?

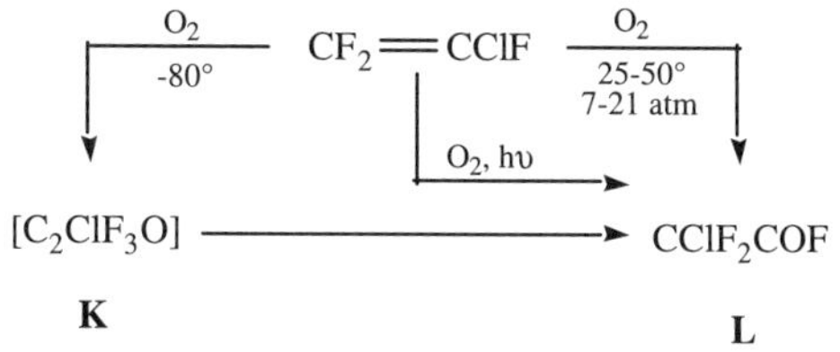

$$CF_2 {=\!\!=} CClF$$

$$[C_2ClF_3O] \longrightarrow CClF_2COF$$

$$\mathbf{K} \qquad\qquad\qquad \mathbf{L}$$

SURPRISE 10

Reaction of Perfluoro-*o*-phenylenediamine with Lead Tetraacetate

Tetrafluoro-*o*-phenylenediamine treated with lead tetraacetate gives a straight-chain compound $C_6F_4N_2$. What is its structure?

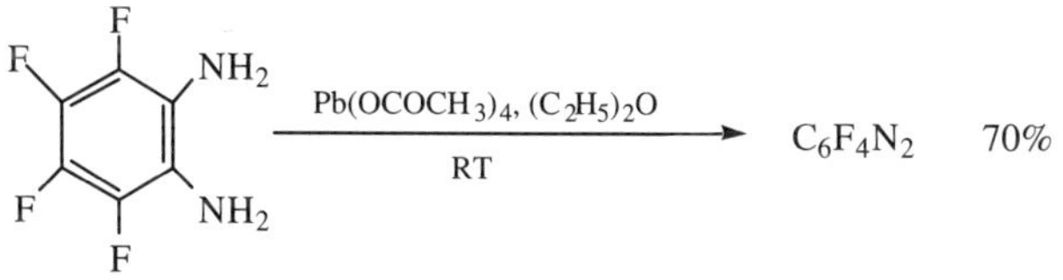

$$\xrightarrow[RT]{Pb(OCOCH_3)_4,\ (C_2H_5)_2O} C_6F_4N_2 \quad 70\%$$

SURPRISE 11

Reaction of Fluorinated Aromatic Diamines with Nitrous Acid

Treatment of 1,3-diamino-4-nitro-2,5,6-trifluorobenzene with nitrous acid gives a compound of empirical formula $C_6FN_5O_4$. What is its structure and how is it formed?

$$\xrightarrow[\text{−5 to 0°}]{\text{2 NaNO}_2,\ \text{H}_2\text{SO}_4} \quad C_6FN_5O_4 \quad 69\%$$

| 5 | **Preparation of Halogen Derivatives** |

SURPRISE 12

Addition of Iodine Fluoride to Chlorotrifluoroethylene

When chlorotrifluoroethylene is treated with iodine pentafluoride and two equivalents of iodine, the iodine fluoride generated adds across the double bond. In contrast to the expectation of a unidirectional addition, two products, **M** and **N**, are formed. What are they?

$$CClF{=}CF_2\ +\ IF \xrightarrow[20°]{\text{IF}_5,\ 2\,\text{I}_2}\ \underset{37\%}{\mathbf{M}}\ +\ \underset{45\%}{\mathbf{N}}$$

SURPRISE 13

Addition of Hydrogen Bromide to Chlorotrifluoroethylene

Addition of hydrogen bromide to chlorotrifluoroethylene may take place in two directions. Bromine may become attached to the carbon linked to two fluorines, or to the carbon with chlorine and fluorine. What is the product of the reaction?

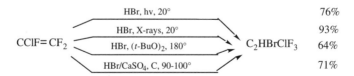

HBr, hv, 20°	76%
HBr, X-rays, 20°	93%
HBr, (t-BuO)$_2$, 180°	64%
HBr/CaSO$_4$, C, 90-100°	71%

$CClF{=}CF_2 \longrightarrow C_2HBrClF_3$

SURPRISE 14

Addition of Hydrogen Chloride to 3,3,3-Trifluoropropene

Under forcing conditions, 3,3,3-trifluoropropene reacts with anhydrous hydrogen chloride. Only one product is formed. What is it?

$$CH_2 = CHCF_3 \; + \; HCl \; \xrightarrow[100°, 9\,h]{AlCl_3} \; C_3H_4ClF_3$$

SURPRISE 15

Addition of Hydrogen Halides to Perfluoropropene

Under strenuous conditions, hydrogen fluoride adds across the double bond of perfluoropropene to form a single reaction product. What is its structure? Hydrogen chloride and hydrogen bromide react similarly.

$$CF_2 = CFCF_3 \; \xrightarrow[230°]{HCl} \; C_3HClF_6 \quad 60\%$$

SURPRISE 16

Reaction of Perfluoroalkenes with Metal Fluorides

What is the product of the reaction between perfluoropropene and potassium fluoride in formamide?

$$CF_2 = CFCF_3 \; \xrightarrow[25°]{KF,\ HCONH_2} \; C_3HF_7$$

SURPRISE 17

Chlorination of *o*-Fluorotoluene

Iron-catalyzed chlorination of *o*-fluorotoluene gives two main products **O** and **P**, (both C_7H_6ClF). What are their structures?

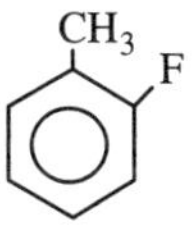

$$\xrightarrow[5\text{-}6\,h]{Cl_2/Fe} \; C_7H_6ClF \; + \; C_7H_6ClF$$
$$\qquad\qquad\qquad \mathbf{O} \qquad\qquad \mathbf{P}$$

SURPRISE 18

Chlorination of 3,4-Difluoronitrobenzene

The product of high-temperature chlorination of 3,4-difluoronitroben-zene is compound **Q**. What are its formula and structure?

$$\text{3,4-difluoronitrobenzene} \xrightarrow[400°]{Cl_2} \mathbf{Q} \quad 82\%$$

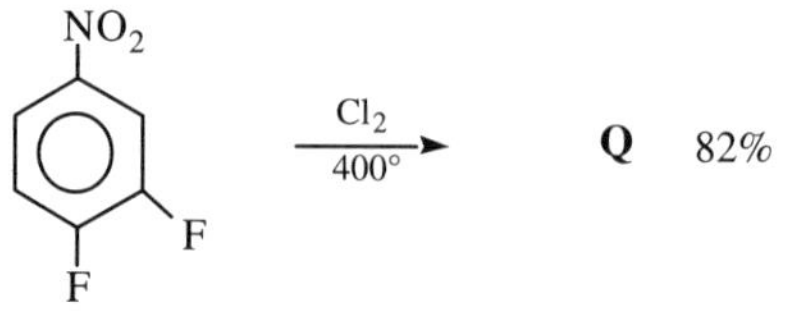

SURPRISE 19

Reaction of Fluoropentanitrobenzene with Hydrogen Chloride

When fluoropentanitrobenzene is treated with hydrogen chloride, compound **R** is formed. What are its formula and structure?

$$\xrightarrow[RT]{HCl, C_6H_6} \mathbf{R}$$

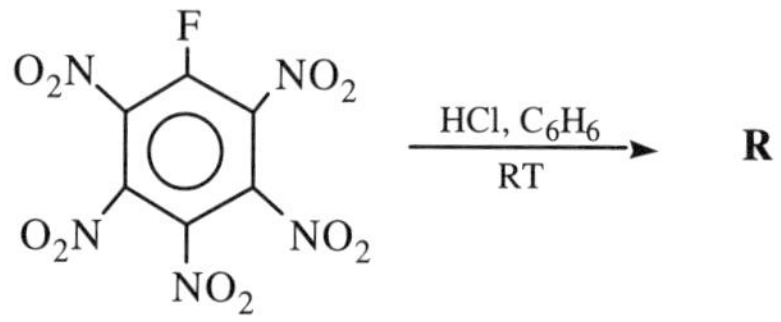

SURPRISE 20

Reaction of 1,1-Difluoroethylene with Trifluoroacetyl Hypochlorite

1,1-Difluoroethylene is treated with trifluoroacetyl hypochlorite to yield a single product. What is its structure?

$$CF_2{=}CH_2 \;+\; CF_3COOCl \xrightarrow{\text{-150° to 22°}} C_4H_2ClF_5O_2$$

SURPRISE 21

Reaction of Pentafluorophenol with *tert*-Butyl Hypobromite

Pentafluorophenol is treated with *tert*-butyl hypobromite to give a compound C_6BrF_5O. What is its structure?

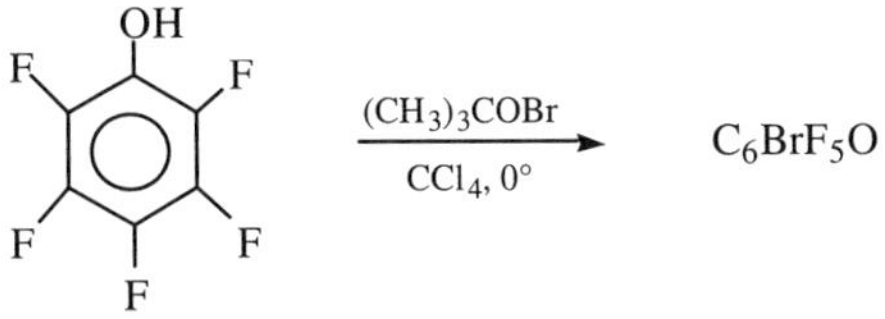

SURPRISE 22

Reaction of Perfluorocycloalkenes with Aluminum Halides

What results from a treatment of perfluorocyclobutene with anhydrous aluminum chloride (compound **S**) or aluminum bromide (compound **T**)?

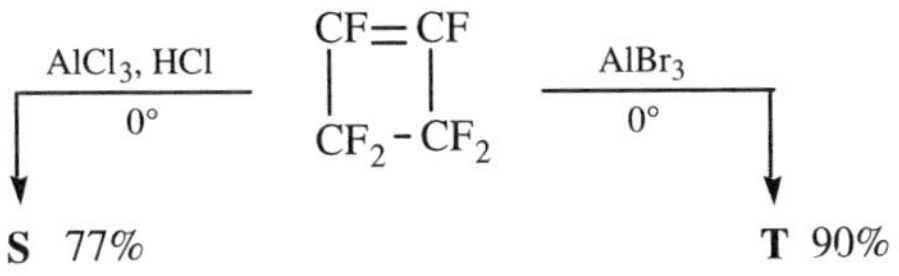

SURPRISE 23

Reaction of Perfluoro-2-butyltetrahydrofuran with Aluminum Chloride

What happens when perfluoro-2-butyltetrahydrofuran is heated with aluminum chloride?

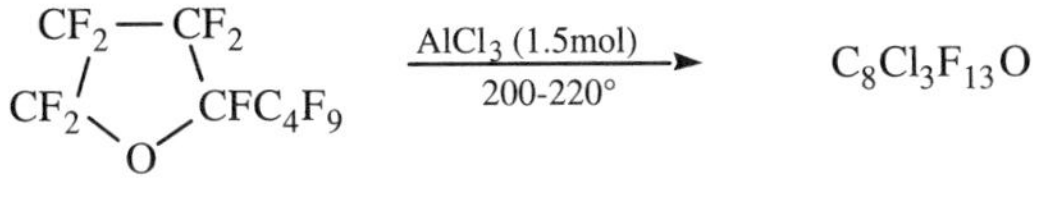

SURPRISE 24

Reaction of Perfluorotetrahydropyran with Aluminum Chloride

What is the product of the reaction of perfluorotetrahydropyran with aluminum chloride?

$$
\begin{array}{c}
\text{CF}_2 \\
\text{CF}_2 \quad \text{CF}_2 \\
\text{CF}_2 \quad \text{CF}_2 \\
\text{O}
\end{array}
\quad
\xrightarrow[180°]{\text{AlCl}_3}
\quad
\text{C}_5\text{Cl}_4\text{F}_6\text{O}
$$

6   Nitration

SURPRISE 25

Nitration of *o*-Fluorotoluene

When *o*-fluorotoluene is nitrated, the main product, obtained with an 84% yield (**U**), is accompanied by two minor byproducts (**V** and **W**). What are the structures of the products **U**, **V**, and **W**?

$$
\begin{array}{c}
\text{CH}_3 \\
\diagdown \text{F}
\end{array}
\quad
\xrightarrow{\text{HNO}_3\ (100\%)}
\quad
\textbf{U} + \textbf{V} + \textbf{W}
$$

7   Reactions of Sulfur Trioxide

SURPRISE 26

Reaction of Sulfur Trioxide with Enol Ethers

Ethyl pentafluoroisopropenyl ether and one equivalent of sulfur trioxide give compound $C_5H_5F_5O_4S$, which on treatment with 2.5 equivalents of trifluoroacetic acid affords compounds $C_3HF_5O_4S$ and $C_4H_5F_3O_2$. What are the compounds shown, and how are they formed?

$$
\begin{array}{c}
\text{OCH}_2\text{CH}_3 \\
\mid \\
\text{CF}_3\text{C}=\text{CF}_2 + \text{SO}_3
\end{array}
\xrightarrow[\rightarrow 30°\text{-}50°]{-15°,\ 20\ \text{min}}
\text{C}_5\text{H}_5\text{F}_5\text{O}_4\text{S}
\xrightarrow[\text{RT}]{2.5\ \text{CF}_3\text{CO}_2\text{H}}
\begin{array}{c}
\text{C}_3\text{HF}_5\text{O}_4\text{S} \\
+ \\
\text{C}_4\text{H}_5\text{F}_3\text{O}_2
\end{array}
$$

SURPRISE 27

Reaction of Sulfur Trioxide with Perfluoro-*tert*-butylacetylene

What is the structure of the product of the reaction of sulfur trioxide with perfluoro-*tert*-butylacetylene?

$$(CF_3)_3CC{\equiv}CF \ + \ SO_3 \quad \xrightarrow{\text{80°, 1 h}} \quad C_6F_{10}O_3S$$

SURPRISE 28

Reaction of Sulfur Trioxide with Perfluoroisobutylene

Perfluoroisobutylene reacts with an excess of sulfur trioxide to yield compounds **X, Y, Z,** and **A**. What are they?

$$(CF_3)_2C{=}CF_2 \ + \ SO_3 \ (\text{excess}) \quad \xrightarrow{\text{170-190°}} \quad \mathbf{X} \ + \ \mathbf{Y} \ + \ \mathbf{Z} \ + \ \mathbf{A}$$

SURPRISE 29

Reaction of Sulfur Trioxide with Perfluoropropylene Oxide

Perfluoropropylene oxide reacts with sulfur trioxide and gives two compounds, **B** and **C**. What are their structures?

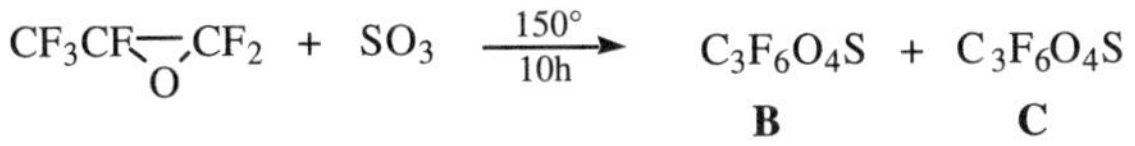

$$CF_3CF{-}CF_2 \ + \ SO_3 \quad \xrightarrow[\text{10h}]{\text{150°}} \quad C_3F_6O_4S \ + \ C_3F_6O_4S$$
$$\overset{}{O} \qquad\qquad\qquad\qquad\qquad \mathbf{B} \qquad\qquad \mathbf{C}$$

SURPRISE 30

Reaction of Sulfur Trioxide with Perfluorotoluene

Perfluorotoluene and sulfur trioxide give a compound $C_7F_8O_3S$. What is its structure?

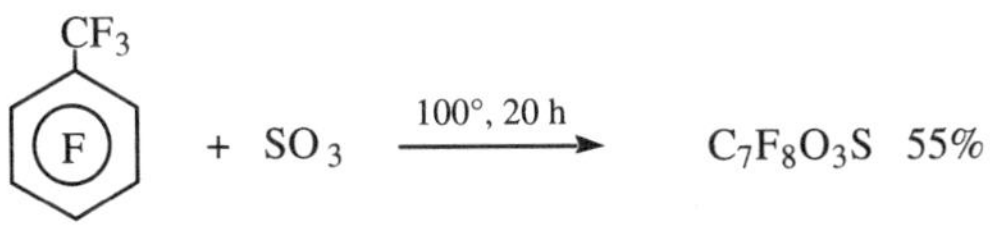

$$\underset{F}{\overset{CF_3}{\bigcirc}} \ + \ SO_3 \quad \xrightarrow{\text{100°, 20 h}} \quad C_7F_8O_3S \ \ 55\%$$

8 Acid-Catalyzed Additions and Substitutions

SURPRISE 31

Reaction of Benzene with 1-Chloro-2-fluoropropane

How does benzene react with 1-chloro-2-fluoropropane under the catalysis by strong Lewis acids such as boron trifluoride or aluminum chloride? What is compound **D**?

$$C_6H_6 \ + \ CH_2ClCHFCH_3 \ \xrightarrow[-10° \text{ to } 10°]{BF_3} \ \textbf{D} \quad 92\%$$

SURPRISE 32

Reaction of Benzene with 2-Chloro-1,1,1-trifluoropropane

How does benzene react with 2-chloro-1,1,1-trifluoropropane in the Friedel-Crafts reaction, and what is the product **E**?

$$C_6H_6 \ + \ CF_3CHClCH_3 \ \xrightarrow[20°, \, 72 \text{ h}]{AlCl_3} \ \textbf{E} \quad 10\%$$

SURPRISE 33

Reaction of Benzene with 3,3,3-Trifluoropropene

What is the product **F** of treatment of benzene and 3,3,3-trifluoropropene with boron trifluoride and hydrogen fluoride?

$$C_6H_6 \ + \ CH_2{=}CHCF_3 \ \xrightarrow[25, \, 4.5 \text{ h}]{HF, \, BF_3} \ \textbf{F} \quad 59\%$$

SURPRISE 34

Reaction of 1-Phenylperfluoropropene with Aluminum Chloride

What is the product of a reaction of 1-phenylperfluoropropene with aluminum chloride?

$$C_6H_5\overset{\displaystyle CF_3}{\underset{\displaystyle F}{\overset{|}{\underset{|}{C}}}}=C\overset{CF_3}{\underset{F}{}} \xrightarrow[-15°,\ 60°]{AlCl_3,\ CH_3COCl} \quad C_9H_4Cl_3F \quad 59\%$$

9   Hydrolyses

SURPRISE 35

Hydrolysis of Benzyl Halides

Why is benzyl fluoride in alkaline or neutral media hydrolyzed more slowly than benzyl chloride and yet in acidic media is hydrolyzed much faster than benzyl chloride?

$$C_6H_5CH_2X \xrightarrow{H_2O} C_6H_5CH_2OH + HX$$

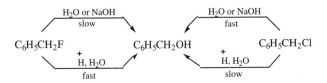

SURPRISE 36

Hydrolysis of ω-Fluorocarboxylic Acids and Their Nitriles

Why is there a difference in the rates of hydrolysis by sodium hydroxide of ω-fluorocarboxylic acids and their nitriles? Explain the results.

$F(CH_2)_3CN$	1 eq. NaOH	205	4
$F(CH_2)_3CO_2H$	1 eq. NaOH	205	74.4
$F(CH_2)_4CN$	4 eq. NaOH	397	0
$F(CH_2)_4CO_2H$	1 eq. NaOH	401	29.6

SURPRISE 37

Hydrolysis of Perfluorocycloalkenes

What are the products of treatment of perfluorocyclopentene with potassium hydroxide in *tert*-butyl alcohol (compound **G**), and in diglyme (compound **H**)?

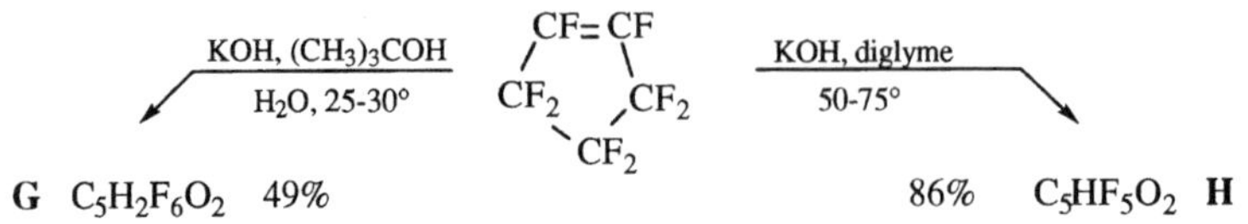

G $C_5H_2F_6O_2$ 49%   86% $C_5HF_5O_2$ **H**

SURPRISE 38

Reaction of 3,5-Dichlorotrifluoropyridine with Potassium Hydroxide

Two products, **I** and **J**, were isolated after treatment of 3,5-dichlorotrifluoropyridine with potassium hydroxide. What are their structures?

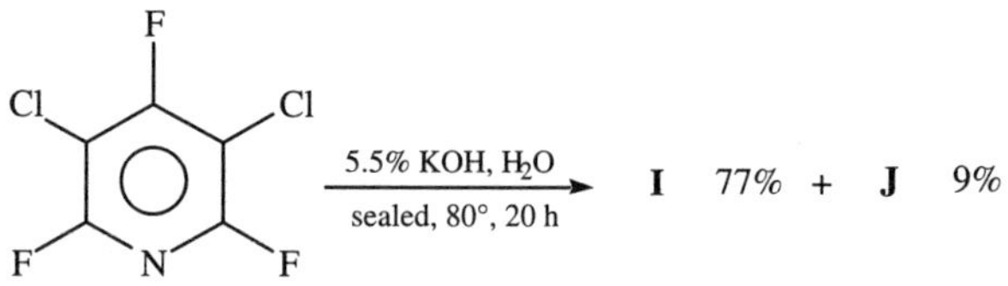

SURPRISE 39

Reaction of 2,5-bis(Trifluoromethyl)aniline with Sodium Hydroxide

What is the result of a reaction of 2,5-bis(trifluoromethyl)aniline with sodium hydroxide (compound **K**)?

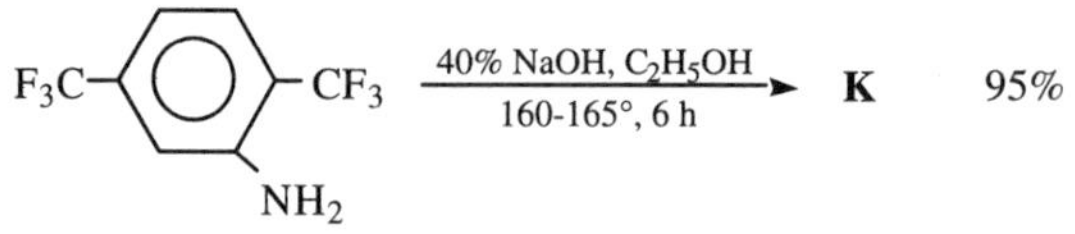

SURPRISE 40

Reaction of Heptafluorobutyraldehyde Hydrate with Potassium Hydroxide

When heptafluorobutyraldehyde hydrate is heated with dilute potassium hydroxide, compound **L** is formed. What is its structure?

$$C_3F_7CH(OH)_2 \xrightarrow[\text{reflux 1.5 h}]{\text{10\% KOH}} \textbf{L} \quad 89\%$$

SURPRISE 41

Reaction of Pentafluoroethyl Iodide with Potassium Hydroxide and Water

What are the compounds resulting from treatment of pentafluoroethyl iodide with potassium hydroxide? Explain the result.

$$CF_3CF_2I \xrightarrow{\text{KOH}} \textbf{M} \quad 74\%$$

SURPRISE 42

Hydrolysis of Unsaturated Fluorinated Compounds

What is the compound resulting from heating 1-(*o*-methoxyphenyl)perfluoropropene with hydrobromic and acetic acid? The reaction product contains phenolic and carboxylic hydroxyls. What is a probable mechanism of the reaction?

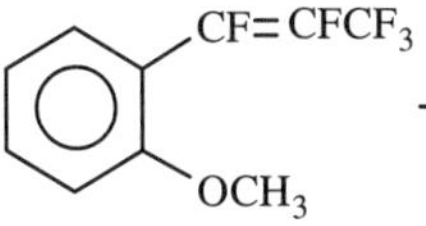

$$\text{(aryl) } \xrightarrow[\text{CH}_3\text{CO}_2\text{H, 140°, 30 h}]{\text{40\% HBr, H}_2\text{O}} C_9H_6F_4O_2 \quad 86\%$$

with substituents CF=CFCF$_3$ and OCH$_3$ on the benzene ring.

SURPRISE 43

Hydrolysis of Geminal Difluorides

What is the product of hydrolysis of the compound shown below?

$$C_6F_{13}CF_2CHFCO_2C_7H_{15} \xrightarrow[\text{2. HCl, H}_2\text{O}]{\text{1. piperidine, CH}_2\text{Cl}_2\text{, boil}} C_{16}H_{18}F_{14}O_4 \ (85\%)$$

SURPRISE 44

Hydrolysis of Difluoromethylene Group in Enol Ethers

Sulfuric acid converts 3,3-difluoro-2-ethoxy-1-phenyl-1-cyclobutene to a fluorine-free product. What is the compound **N** and how is it formed?

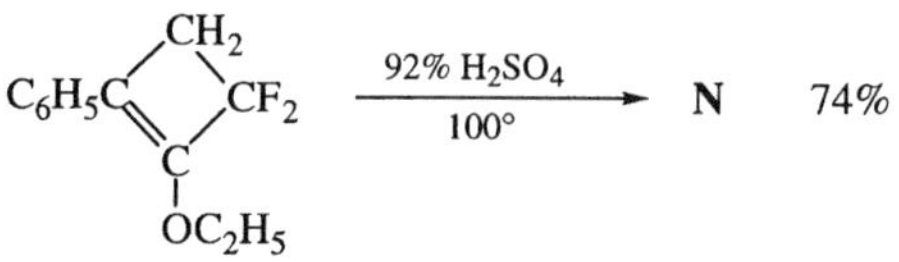

$$C_6H_5C \overset{CH_2}{\underset{\underset{OC_2H_5}{C}}{\diagup\diagdown}} CF_2 \quad \xrightarrow[100°]{92\% \; H_2SO_4} \quad \textbf{N} \quad 74\%$$

SURPRISE 45

Reaction of Alkalis with Fluorinated Cyclopropanes

When 1-acetyl-2,2-difluoro-3-phenylcyclopropane is heated with potassium hydroxide and methanol, the reaction product is a keto ester. How is it formed, and what is its structure?

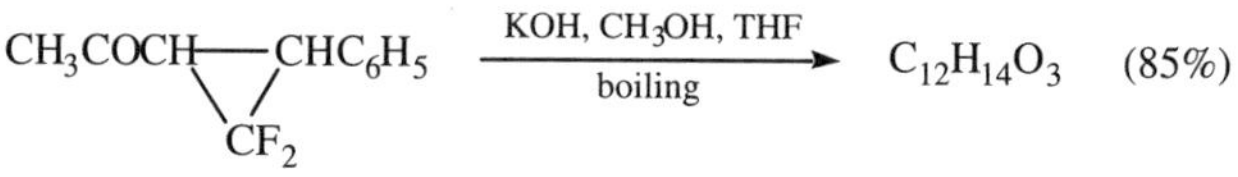

$$CH_3COCH \overset{\diagup\diagdown}{\underset{CF_2}{\diagdown\diagup}} CHC_6H_5 \quad \xrightarrow[\text{boiling}]{KOH, \; CH_3OH, \; THF} \quad C_{12}H_{14}O_3 \quad (85\%)$$

SURPRISE 46

Product of Treatment of *p*-Trifluoromethylphenol with Potassium Hydroxide

p-Trifluoromethylphenol treated with dilute methanolic potassium hydroxide under relatively mild conditions gives *p*-hydroxybenzoic acid. Why does the hydrolysis of the trifluoromethyl group occur under mild conditions, and what is a probable mechanism of the reaction?

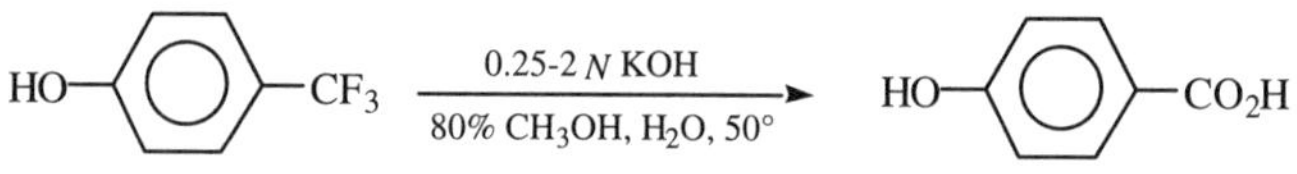

$$HO-\!\!\left\langle \bigcirc \right\rangle\!\!-CF_3 \quad \xrightarrow[80\% \; CH_3OH, \; H_2O, \; 50°]{0.25\text{-}2 \, N \; KOH} \quad HO-\!\!\left\langle \bigcirc \right\rangle\!\!-CO_2H$$

10 Alkylations

SURPRISE 47

Reaction of 1-Chloro-3,3,4,4-tetrafluorocyclobutene with Potassium Hydroxide and Ethanol

What are the products of treatment of 1-chloro-3,3,4,4-tetrafluoro-cyclobutene with an excess of potassium hydroxide in ethanol (compounds **O** and **P**)?

$$\begin{array}{c} CF_2-CH \\ | \quad \| \\ CF_2-CCl \end{array} \xrightarrow[0°]{KOH,\ C_2H_5OH} \mathbf{O} \quad 77\% \xrightarrow[0°]{KOH,\ C_2H_5OH} \mathbf{P} \quad 82\%$$

SURPRISE 48

Reaction of 1-Bromo-2-chlorotetrafluorocyclobutene with Potassium Hydroxide and Ethanol

When 1-bromo-2-chloro-3,3,4,4-tetrafluorocyclobutene is treated with ethanolic potassium hydroxide, compounds **Q** and **R** are formed. What are their structures, and how are they formed?

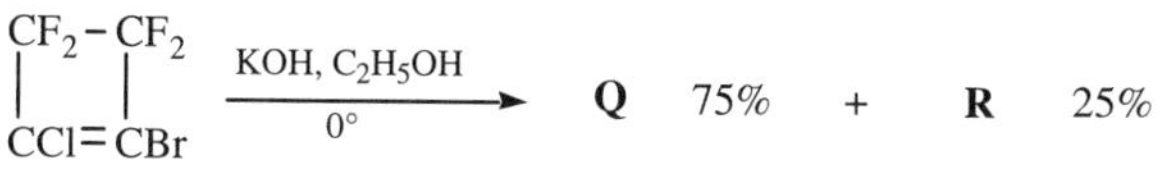

$$\begin{array}{c} CF_2-CF_2 \\ | \quad \quad | \\ CCl=CBr \end{array} \xrightarrow[0°]{KOH,\ C_2H_5OH} \mathbf{Q} \quad 75\% \quad + \quad \mathbf{R} \quad 25\%$$

SURPRISE 49

Reaction between 1-Chloro-2-fluoroethane and Ethyl Acetoacetate

In order to prepare 5-fluoro-2-pentanone by the acetoacetate synthesis, the experimenter alkylated ethyl acetoacetate with 1-chloro-2-fluoro-ethane in the presence of sodium ethoxide in ethanol. Instead of the expected ethyl 3-ethoxycarbonyl-5-fluoro-2-pentanone, he obtained a fluorine-free compound. What was this product, and how was it formed?

$$FCH_2CH_2Cl + CH_3COCH_2CO_2C_2H_5 \xrightarrow[C_2H_5OH]{C_2H_5ONa} C_8H_{12}O_3$$

SURPRISE 50

Reaction of Pentafluorophenol with Chlorodifluoromethane and Sodium Hydroxide

Pentafluorophenol is heated with chlorodifluoromethane and sodium hydroxide in aqueous dioxane. What is product **S**?

$$C_6F_5OH + CHClF_2 \xrightarrow[70°, \ 20 \ min]{NaOH, \ aq. \ dioxane} S \quad 90\%$$

SURPRISE 51

Reaction of 1,4-Dibromohexafluoro-2-butene with Sodium Ethoxide

Treatment of 1,4-dibromohexafluoro-2-butene with sodium ethoxide in ethanol gives a compound of empirical formula $C_6H_5BrF_6O$. What is its structure, and how is it formed?

$$BrCF_2CF=CFCF_2Br \xrightarrow[-10° \ to \ -15°, \ then \ RT, \ 0.5 \ h]{C_2H_5ONa, \ C_2H_5OH} C_6H_5BrF_6O$$

SURPRISE 52

Reaction of Perfluoro-3,4-dimethyl-3-hexene with Methanol and Pyridine

Perfluoro-3,4-dimethyl-3-hexene refluxed with methanol and pyridine gives compound **T**, which is ultimately converted to compound **U**. What are the compounds **T** and **U**?

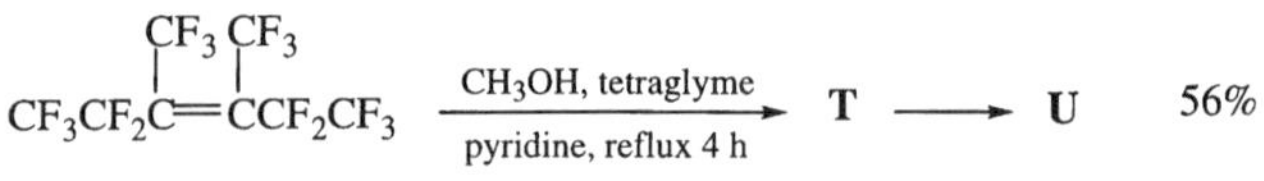

$$CF_3CF_2\overset{\overset{\displaystyle CF_3}{|}}{C}=\overset{\overset{\displaystyle CF_3}{|}}{C}CF_2CF_3 \xrightarrow[pyridine, \ reflux \ 4 \ h]{CH_3OH, \ tetraglyme} T \longrightarrow U \quad 56\%$$

SURPRISE 53

Reaction of Chlorotrifluoroethylene with Sodium Ethoxide

Refluxing chlorotrifluoroethylene with an excess of sodium ethoxide in tetrahydrofuran gives two compounds, **V** and **W**. What are these compounds, and how are they formed?

$$CF_2=CClF \xrightarrow[\text{-20°, 15 min, reflux 24 h}]{\text{3 } C_2H_5ONa, \text{ THF}} \textbf{V} \quad 30\% \ + \ \textbf{W} \quad 20\%$$

SURPRISE 54

Reaction of Perfluoro-γ-butyrolactone with Perfluoropropylene Oxide

What is the product **X** of the reaction of perfluoro-γ-butyrolactone with perfluoropropylene oxide in the presence of cesium fluoride?

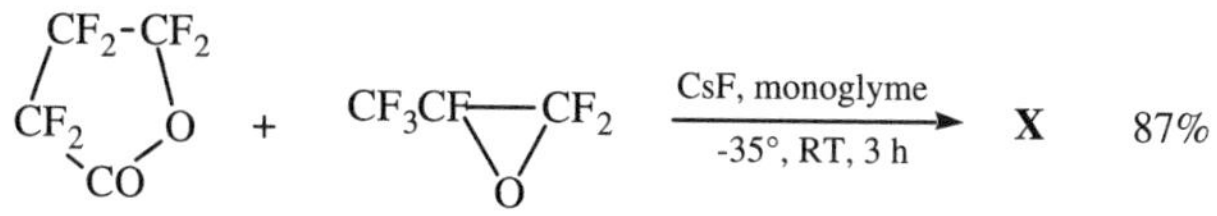

$$\underset{\substack{CF_2-CF_2 \\ CF_2 \quad \ \ O \\ CO}}{} \ + \ \underset{\substack{CF_3CF—CF_2 \\ O}}{} \xrightarrow[\text{-35°, RT, 3 h}]{\text{CsF, monoglyme}} \textbf{X} \quad 87\%$$

SURPRISE 55

Reaction of Alkali Thiophenoxides with Polyfluorohalomethanes

Treatment of potassium thiophenoxide with dichlorodifluoromethane in dimethylformamide (DMF) affords three products, **Y**, **Y′**, and **Y″**. What are compounds **Y**, **Y′**, and **Y″**, and how are they formed?

$$C_6H_5SK \xrightarrow[\text{2. 17\% HCl}]{\text{1. } CCl_2F_2, \text{ DMF, RT, 2.7 atm, 4 h}} \underset{62\% \quad 8\% \quad 7\%}{\textbf{Y} \ + \ \textbf{Y'} \ + \ \textbf{Y''}}$$

SURPRISE 56

Reaction of Perfluorodecalin with Sodium Thiophenoxide

It is hard to believe how the unexpected product octakis-(phenylthio)-naphthalene is formed. Any suggestion?

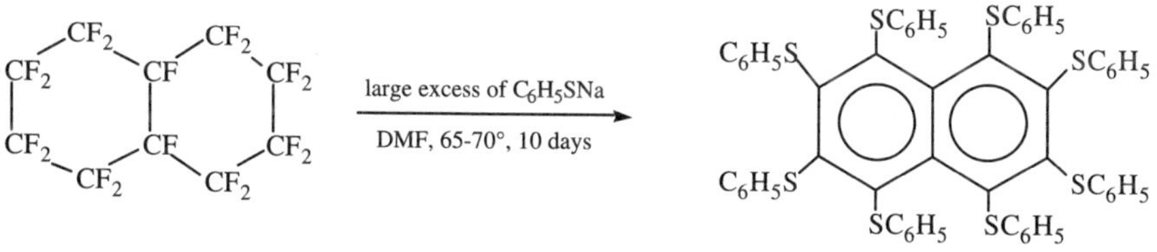

$$\text{(perfluorodecalin)} \xrightarrow[\text{DMF, 65-70°, 10 days}]{\text{large excess of } C_6H_5SNa} \text{(octakis-(phenylthio)naphthalene)}$$

SURPRISE 57

Reaction of Perfluorocyclobutene with Hydrazine

How does perfluorocyclobutene react with a large excess of hydrazine? What is the fluorine-free product **Z**?

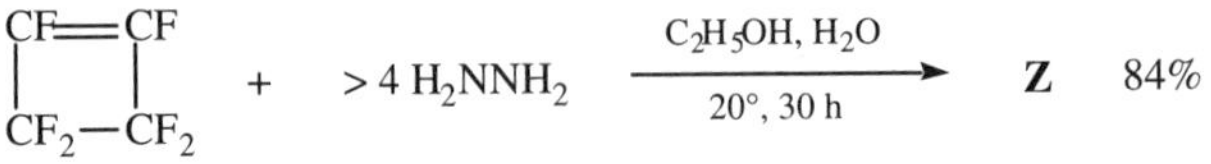

$$\begin{array}{c} CF{=}CF \\ | \qquad | \\ CF_2{-}CF_2 \end{array} \quad + \quad > 4\ H_2NNH_2 \quad \xrightarrow[20°,\ 30\ h]{C_2H_5OH,\ H_2O} \quad \textbf{Z} \quad 84\%$$

SURPRISE 58

Reaction of 1,2,3,5-tetrakis(Trifluoromethyl)benzene with Ammonia

What is the product **A** of treatment of 1,2,3,5-tetrakis(trifluoromethyl)-benzene with an excess of ammonia?

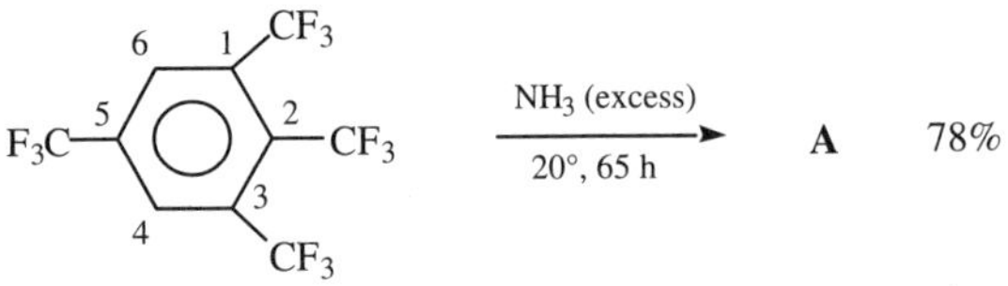

$$\xrightarrow[20°,\ 65\ h]{NH_3\ (excess)} \quad \textbf{A} \quad 78\%$$

SURPRISE 59

Reaction of Perfluoroisobutylene with Dimethyl Malonate

How does perfluoroisobutylene react with dimethyl malonate, and what is product **B**?

$$(CF_3)_2C{=}CF_2 + CH_2(CO_2CH_3)_2 \quad \xrightarrow[(C_2H_5)_2O,\ 20°,\ 5\ days]{BF_3,\ N(C_2H_5)_3} \quad \textbf{B} \quad 45\%$$

SURPRISE 60

Reaction of Perfluorobenzotrichloride and Chlorodifluoromethane

Heating of perfluorobenzotrichloride and chlorodifluoromethane gives a mixture of which the main product is compound **C**. How is it formed?

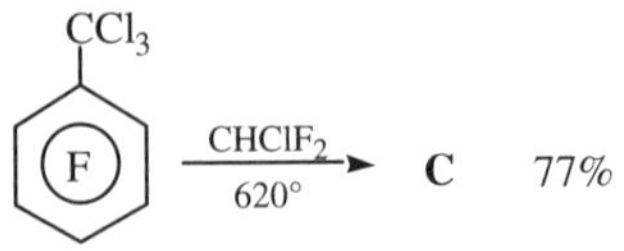

$$\xrightarrow[620°]{CHClF_2} \quad \textbf{C} \quad 77\%$$

SURPRISE 61

Reaction of Enamines with Trifluoromethanesulfenyl Chloride

Enamine of 2-phenylaminocarbonylcyclopentanone and morpholine reacts with two molecules of trifluoromethanesulfenyl chloride. What is product **D**?

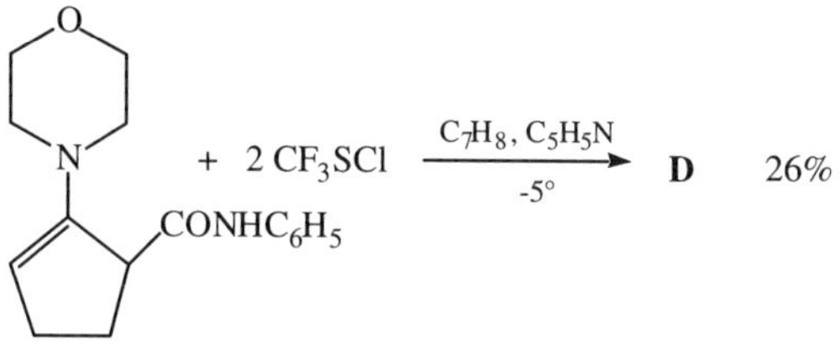

$$\text{(structure)} + 2\ CF_3SCl \xrightarrow[-5°]{C_7H_8,\ C_5H_5N} \mathbf{D} \quad 26\%$$

SURPRISE 62

Reaction of 1,1-Dichlorodifluoroethylene with Methanol

What are products **E** and **F** of the reaction of 1,1-dichloro-2,2-difluoro-ethylene with methanol in the presence of sodium methoxide?

$$CCl_2{=}CF_2 + CH_3OH \xrightarrow{CH_3ONa} \mathbf{E + F} \quad 90\%$$

SURPRISE 63

Reaction of Chlorotrifluoroethylene with Aniline

What is the product of a reaction of chlorotrifluoroethylene with aniline?

$$CClF{=}CF_2 + C_6H_5NH_2 \text{ (excess)} \xrightarrow[720\ h]{22\text{-}29°} \mathbf{G} \quad 63\%$$

SURPRISE 64

Reaction of Chlorotrifluoroethylene with Dimethylamine

When chlorotrifluoroethylene is heated with four molecules of dimethyl-amine, what reaction product is formed?

$$CClF{=}CF_2 + 7.5\ (CH_3)_2NH \xrightarrow[8\ h]{58\text{-}62°} \mathbf{H} \quad 54\%$$

SURPRISE 65

Reaction of Alcohols with Perfluoroalkenes

Surprise 62 addressed nucleophilic additions to fluorinated alkenes. Here, a free-radical addition is presented. What is the product of a reaction of methanol and 1-perfluorobutene in the presence of dibenzoyl peroxide?

$$CF_2{=}CFCF_2CF_3 + CH_3OH \xrightarrow[110\text{-}120°]{(C_6H_5COO)_2} \quad \textbf{I} \quad 76\%$$

SURPRISE 66

Reaction of Fluoroalkenes with Azides

What are the structures of the products of reaction of 1H-pentafluoropropene with sodium azide or triethylammonium azide?

$$CHF{=}CFCF_3 \xrightarrow{\overline{N}_3} C_3H_2F_5N_3 + C_3HF_4N_3$$

11 Arylations

SURPRISE 67

Reaction of 1,2-Difluorotetrachlorobenzene with Sodium Methoxide

What happens when 1,2-difluorotetrachlorobenzene is refluxed with sodium methoxide in methanol? What product (**J**) is formed?

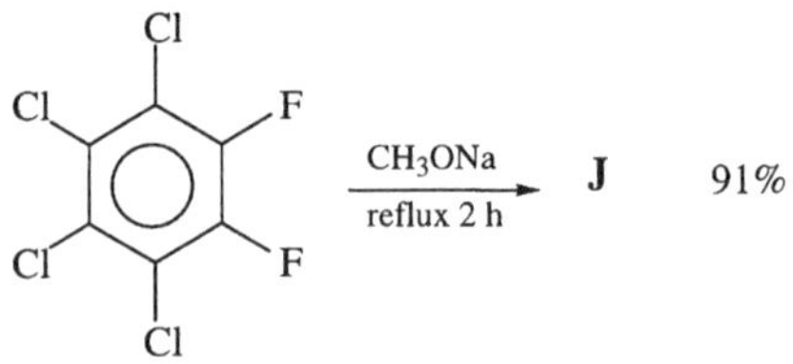

$$\xrightarrow[\text{reflux 2 h}]{CH_3ONa} \quad \textbf{J} \quad 91\%$$

SURPRISE 68

Reaction of 2,4-Dinitrohalobenzenes with Sodium Methoxide

In the reaction with nucleophiles, halobenzenes with electron-withdrawing substituents such as nitro group in *ortho* or *para* (or both) positions activate the halogens for nucleophilic displacement. Which halogen is replaced more easily, chlorine or fluorine? Explain why.

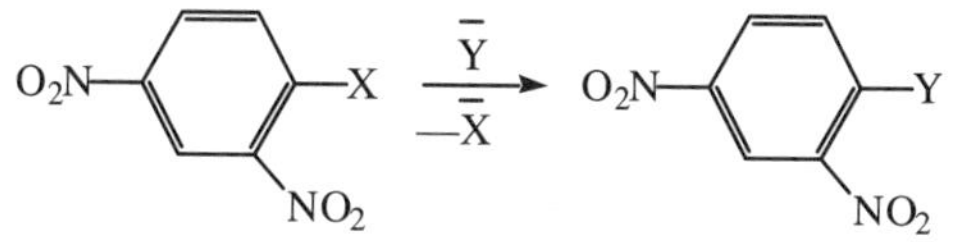

SURPRISE 69

Reaction of Fluoropentachlorobenzene with Potassium *p*-Methoxyphenoxide

In the presence of 18-crown-6 ether, which complexes potassium ion, fluoropentachlorobenzene reacts with potassium *p*-methoxyphenoxide to give compound **K**. What is compound **K**?

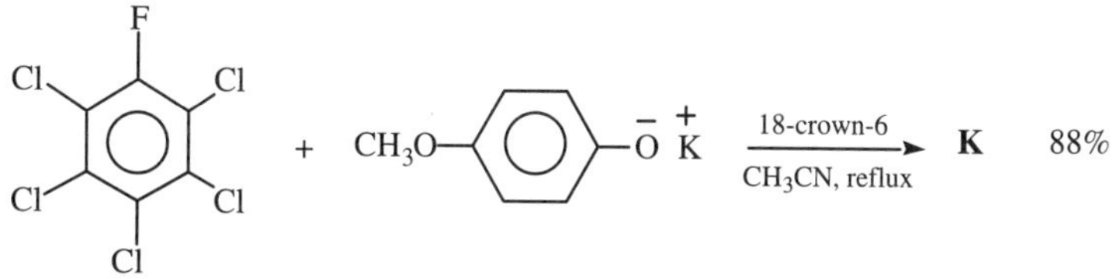

SURPRISE 70

Reaction of Perfluoronaphthalene with Sodium Alkoxides or Aryloxides

What happens when perfluoronaphthalene is heated with sodium *m*-methylphenoxide? What is reaction product **L**?

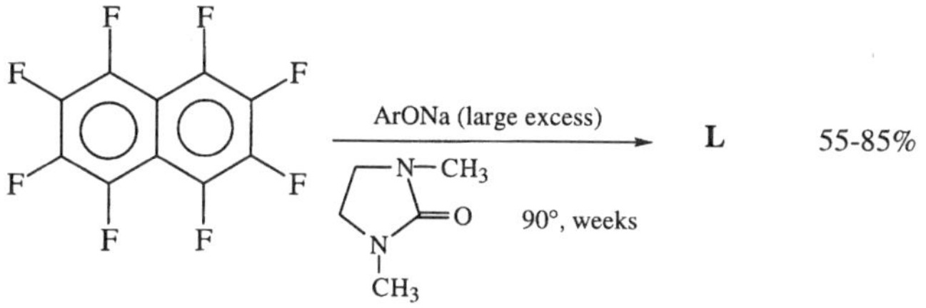

Reaction of Perfluoropyridine with Methanol

What reaction takes place when a mixture of perfluoropyridine and methanol is irradiated with ultraviolet light? What compound **M** is formed?

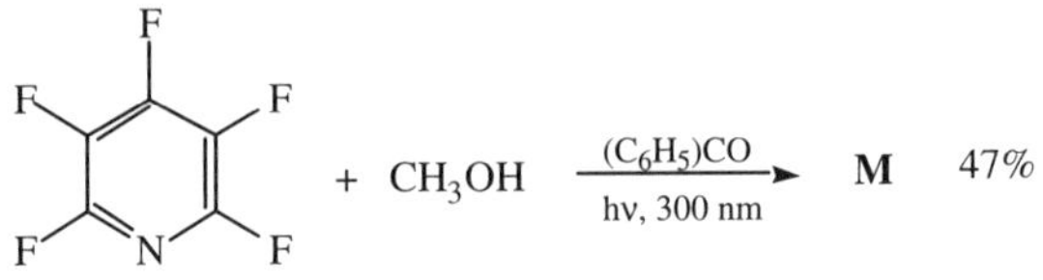

$$\text{(perfluoropyridine)} + \text{CH}_3\text{OH} \xrightarrow[\text{hv, 300 nm}]{(\text{C}_6\text{H}_5)\text{CO}} \textbf{M} \quad 47\%$$

Arylation at Sulfur

When 2-bromo-4,5-difluoronitrobenzene is heated with sodium sulfide, a diaryl sulfide is formed. The nitro group activates halogens in *ortho-* and *para* positions for nucleophilic displacements. Which of the activated halogens is replaced in the reaction with sodium sulfide giving a diaryl sulfide, and what is the product?

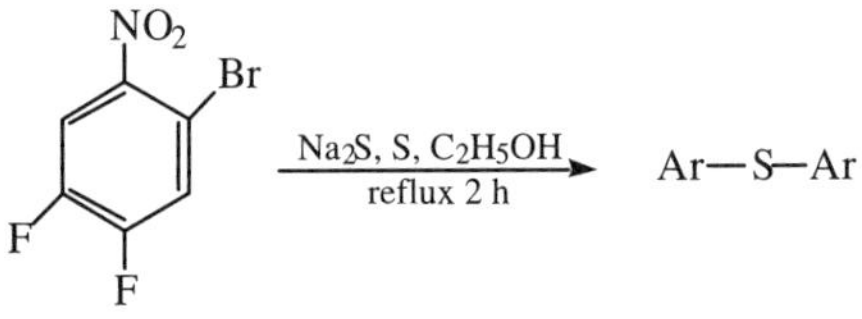

$$\xrightarrow[\text{reflux 2 h}]{\text{Na}_2\text{S, S, C}_2\text{H}_5\text{OH}} \quad \text{Ar}-\text{S}-\text{Ar}$$

---

12 Acylations

Acylation with Mixed Anhydrides of Trifluoroacetic Acid

When carboxylic acids are treated with trifluoroacetic anhydride, mixed anhydrides (acyl trifluoroacetates) are formed. These compounds are useful acylating agents. What is the product of the reaction of an alcohol with acyl trifluoroacetate?

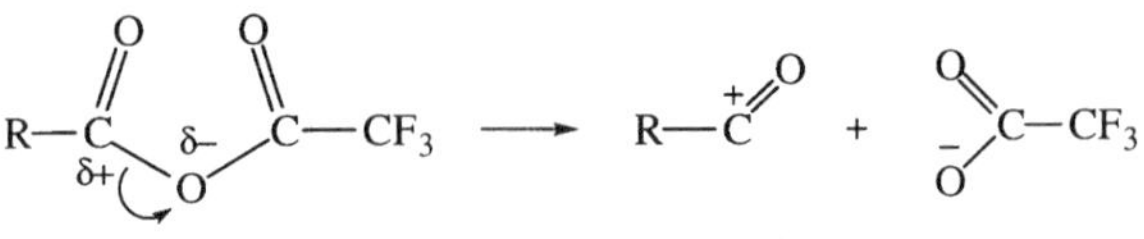

13 Aldol-Type Condensations

SURPRISE 74

Reaction of 1,1,1-Trifluoroacetone, Formaldehyde, and Piperidine

How do 1,1,1-trifluoroacetone and formaldehyde react with piperidine, and what is the structure of product **N**?

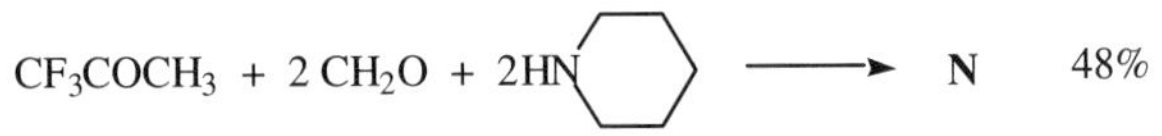

$$CF_3COCH_3 + 2\,CH_2O + 2HN\!\!\bigcirc \longrightarrow N \quad 48\%$$

SURPRISE 75

Reaction of α,α,α-Trifluoroacetophenone with Tributylphosphine

α,α,α-Trifluoroacetophenone and tributylphosphine react to form two geometric isomers, **O** and **P**. What are their structures?

$$C_6H_5COCF_3 + (C_4H_9)_3P \xrightarrow[\text{reflux 20 h}]{C_6H_{14}} O\ 11\% + P\ 32\%$$

SURPRISE 76

Reaction of Dibromodifluoromethane and tris(Dimethylamino)phosphine with Fluorinated Ketones

When dibromodifluoromethane and tris(dimethylamino)phosphine (hexamethylphosphorous amide) react with fluorinated ketones, the oxygen-free product **Q** is obtained. What is it, and how is it formed?

$$CBr_2F_2 + 2\,[(CH_3)_2N]_3P \xrightarrow{C_6H_5COCF_3} Q \quad 85\%$$

SURPRISE 77

Reactions of Phosphorus Ylides with Fluorinated Nitriles

Ethoxycarbonylmethenyltriphenylphosphorane reacts with fluorinated nitriles to form intermediate **R**, which is hydrolyzed to the fluorinated β-ketoester **S**. Explain the sequence of reactions leading to these products.

$$(C_6H_5)_3P=CHCO_2C_2H_5 \ + \ CF_3\,CN \ \longrightarrow \ R \ \xrightarrow[H^+]{H_2O} \ S \quad 90\%$$

SURPRISE 78

Reaction of tris(Trifluoromethyl)methane with Acrylonitrile

What is the product **T** of the reaction of tris(trifluoromethyl)methane with acrylonitrile in the presence of triethylamine?

$$(CF_3)_3CH \ + \ CH_2=CHCN \ \xrightarrow[\substack{Sealed \\ 100°,\ 77\ h}]{(C_2H_5)_3N} \ T \quad 66\%$$

---

14 Organometallic Syntheses

SURPRISE 79

Reaction of Ethyl Chlorofluoroacetate with Grignard Reagents

When ethyl chlorofluoroacetate is treated with phenylmagnesium bromide, compound **U** is formed. This compound reacts with methylmagnesium bromide to give compound **V**. Refluxing of the product in tetrahydrofuran affords compound **W**. What are compounds **U**, **V**, and **W**, and how are they formed?

$$CHClFCO_2C_2H_5 \ \xrightarrow[\substack{-60° \\ 55\%}]{C_6H_5MgBr} \ U \ \xrightarrow[-40°]{CH_3MgBr} \ V \ \xrightarrow[reflux]{THF} \ W$$

SURPRISE 80

Reaction of 3-Chloropentafluoropropene with Phenylmagnesium Bromide

Phenylmagnesium bromide reacts with 3-chloropentafluoropropene to give an intermediate **X**, which decomposes to two products, **Y**, and **Z**. What are compounds **X**, **Y**, and **Z**, and how are they formed?

$$CF_2{=}CFCClF_2 + C_6H_5MgBr \longrightarrow X \diagdown$$

Y	9%
Z	33%

SURPRISE 81

Reaction of 1,2-Dichlorohexafluorocyclopentene with Ethylmagnesium Bromide

When 1,2-dichlorohexafluorocyclopentene is treated with an excess of ethylmagnesium bromide, two products, **A** and **B**, are obtained. What are they, and how are they formed?

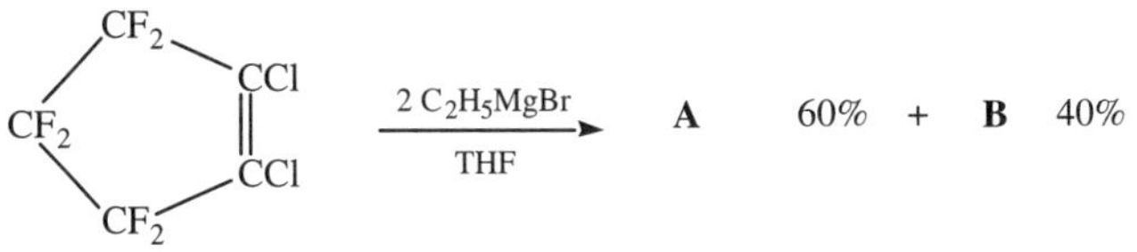

$$\xrightarrow[\text{THF}]{2\ C_2H_5MgBr} \quad A \quad 60\% \ + \ B \quad 40\%$$

SURPRISE 82

Reaction of 1,2-Dichloro-1-fluoroethylene with Butyllithium and Acetone

When 1,2-dichloro-1-fluoroethylene is consecutively treated with butyllithium and acetone, a tertiary alcohol of the structure shown below is obtained as the product. What is **X** in the formula shown?

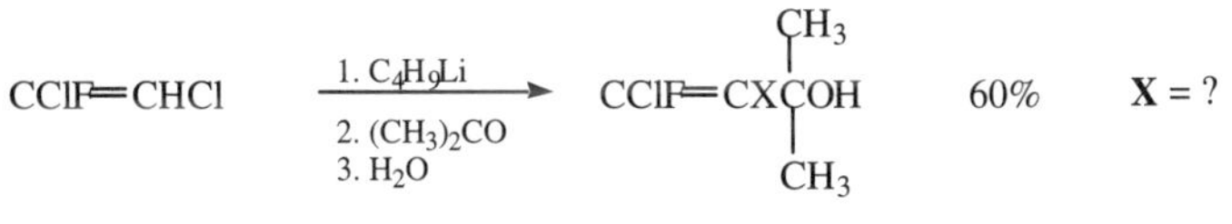

$$CClF{=}CHCl \xrightarrow[\substack{2.\ (CH_3)_2CO \\ 3.\ H_2O}]{1.\ C_4H_9Li} CClF{=}CXCOH \quad 60\% \quad X = ?$$

(with CH_3 groups above and below the central carbon)

SURPRISE 83

Reaction of Trifluoroethylene with Butyllithium and Acetone

Trifluoroethylene or bromotrifluoroethylene is treated with butyllithium to give an organometallic compound **C** which is subsequently treated with acetone. After addition of water, a tertiary alcohol **D** is obtained, which is converted ultimately to the final product **E**.

$$CF_2{=}CHF \xrightarrow{\ C_4H_9Li\ } C \xrightarrow{\ (CH_3)_2CO\ } D \xrightarrow{\ H_2O\ } E \qquad 30\%$$

SURPRISE 84

Reaction of 1-Chloro-1,2-difluoroethylene with Butyllithium and Carbon Dioxide

When 1-chloro-1,2-difluoroethylene is treated consecutively with butyllithium in tetrahydrofuran (THF) and carbon dioxide, two products, **F** and **G**, are obtained after esterification with methanol. What are these products?

$$CClF{=}CHF \atop E/Z = 1{:}1 \qquad \xrightarrow[\text{2. CO}_2 \quad \text{3. CH}_3\text{OH}]{\text{1. C}_4\text{H}_9\text{Li, THF, (C}_2\text{H}_5)_2\text{O, -115}°} \quad F \ + \ G \qquad 1{:}2$$

SURPRISE 85

Reaction of Aldehydes with 1,1,1-Trichlorotrifluoroethane and Zinc in the Presence of Acetic Anhydride

Benzaldehyde, 1,1,1-trichlorotrifluoroethane, and zinc are heated in dimethylformamide (DMF) in the presence of acetic anhydride. What are products **H** and **I**?

$$C_6H_5CHO \ + \ CCl_3CF_3 \ + \ {>}2\ Zn \quad \xrightarrow[\text{50}°,\ 7\ h]{\text{DMF, (CH}_3\text{CO)}_2\text{O}} \quad H\ 65\% \ + \ I\ 11\%$$

SURPRISE 86

Reaction of Aldehydes with 1,1,1-Trichlorotrifluoroethane and Zinc in the Presence of Aluminum Chloride

What is the product **J** of the reaction of benzaldehyde, 1,1,1-trichlorotrifluoroethane, and zinc in the presence of aluminum chloride?

$$C_6H_5CHO + CCl_3CF_3 \xrightarrow[\text{diglyme, 50°, 18 h}]{\text{DMF, AlCl}_3} \quad \textbf{J} \quad 86\%$$

SURPRISE 87

Reaction of Dibromodifluoromethane and 4-Chloro-3-nitrobenzotrifluoride with Copper

What happens when a mixture of dibromodifluoromethane and 4-chloro-3-nitrobenzotrifluoride is heated with copper? What is the product **K**, and how is it formed?

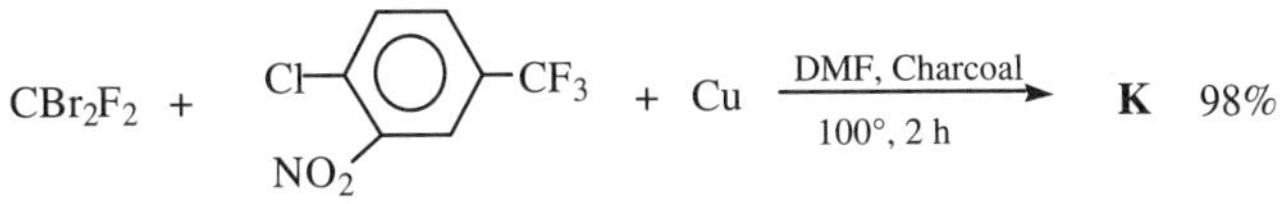

$$CBr_2F_2 + \underset{NO_2}{Cl\!-\!\!\langle\bigcirc\rangle\!\!-\!CF_3} + Cu \xrightarrow[\text{100°, 2 h}]{\text{DMF, Charcoal}} \quad \textbf{K} \quad 98\%$$

15   Additions

SURPRISE 88

Reaction of 2-Aminoethanol with Ethyl 4,4,4-Trifluoro-3-trifluoromethyl-2-butenoate

What is the product **L** of the reaction of 2-aminoethanol with ethyl bis(trifluoromethyl)acrylate (ethyl 4,4,4-trifluoro-3-trifluoromethyl-2-butenoate)? Explain the result.

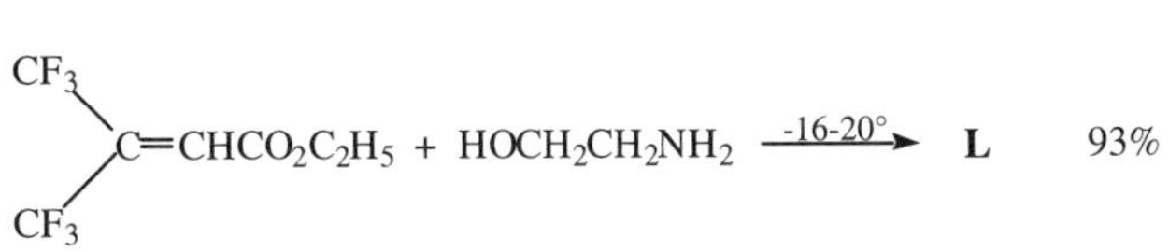

$$\underset{CF_3}{\overset{CF_3}{\diagdown}}C{=}CHCO_2C_2H_5 + HOCH_2CH_2NH_2 \xrightarrow{-16\text{-}20°} \textbf{L} \quad 93\%$$

SURPRISE 89

Reaction of 1,1-Dichlorodifluoroethylene with 1,3-Butadiene

How does 1,1-dichlorodifluoroethylene react with 1,3-butadiene? What product **M** is formed?

$$CCl_2{=}CF_2 + CH_2{=}CH{-}CH{=}CH_2 \longrightarrow \textbf{M}$$

SURPRISE 90

Reaction of 1,2-Dichlorodifluoroethylene with Perfluoropropylene Oxide

What reaction takes place between 1,2-dichlorodifluoroethylene and per-fluoropropylene oxide? What are the products **N** and **O**?

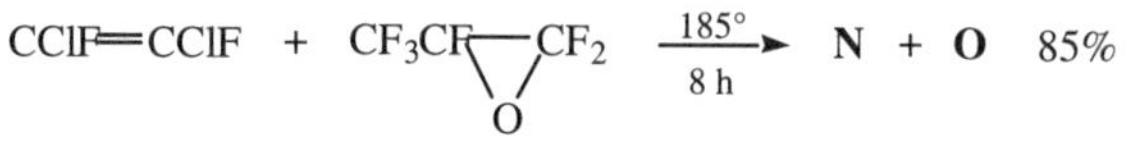

$$CClF{=}CClF \ + \ CF_3CF{-}CF_2 \quad \xrightarrow[8\,h]{185°} \quad N \ + \ O \quad 85\%$$
$$\underset{O}{\diagdown\diagup}$$

SURPRISE 91

Reaction between Perfluorovinylsulfur Pentafluoride and 1,3-Butadiene

How does perfluorovinylsulfur pentafluoride react with 1,3-butadiene? What products **P** and **Q** are formed?

$$CF_2{=}CFSF_5 \ + \ CH_2{=}CH{-}CH{=}CH_2 \quad \xrightarrow[22\,h]{190°} \quad P \ 35\% \ + \ Q \ 35\%$$

SURPRISE 92

Reaction between Trifluoroethylene and 1,3-Butadiene

Trifluoroethylene reacts with 1,3-butadiene to give three compounds, **R**, **S**, and **T**. What are their structures, and how are they formed?

$$CF_2{=}CHF \ + \ CH_2{=}CH{-}CH{=}CH_2 \quad \xrightarrow[24\,h]{215°} \quad \underset{60\% \quad 28\% \quad 13\%}{R \ + \ S \ + \ T}$$

SURPRISE 93

Reaction between Hexafluoroacetone Azine and Acetylene

What is the product $C_{10}H_4F_{12}N_2$ of the reaction of hexafluoroacetone azine with two molecules of acetylene?

$$(CF_3)_2C{=}N{-}N{=}C(CF_3)_2 \ + \ 2\ CH{\equiv}CH \quad \longrightarrow \quad C_{10}H_4F_{12}N_2 \quad 83\%$$

SURPRISE 94

Reaction of tris(*tert*-Butyl)azete and Trifluoroacetonitrile

When tris(*tert*-butyl)azete is treated with trifluoroacetonitrile, a compound of empirical formula $C_{17}H_{27}F_3N_2$ is obtained. What is its structure?

(CH3)3C
C(CH3)3
N
(CH3)3C
+ CF3CN
CH2Cl2
20°
C17H27F3N2
91%

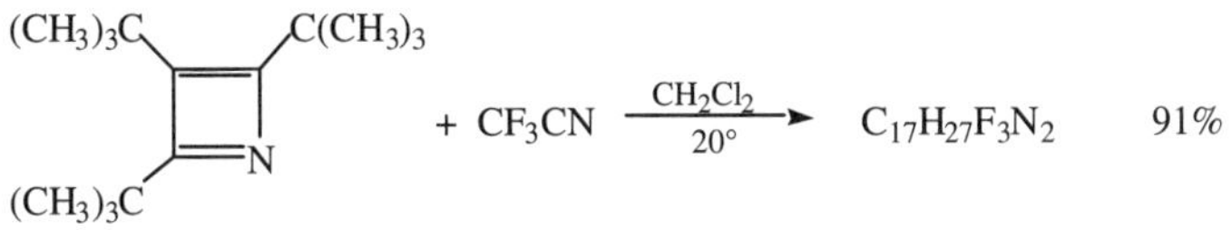

SURPRISE 95

Dimerization of Perfluoro-1,3-butadiene

At 200°C, perfluoro-1,3-butadiene dimerizes. What is the structure of the dimer?

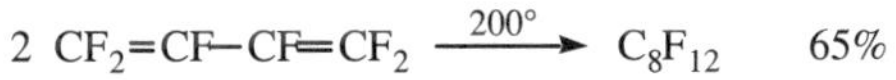

2 CF2=CF–CF=CF2
200°
C8F12
65%

16 Eliminations

SURPRISE 96

Reaction of Alkalis with Fluorinated Cyclopropanes

When 2,2-difluoro-3,3-dimethyl-1-phenylsulfonylmethylcyclopropane is treated with lithium diisopropyl amide (LDA) or potassium hydroxide, a compound of empirical formula $C_{12}H_{13}FO_2S$ is obtained. What is its structure, and how is it formed?

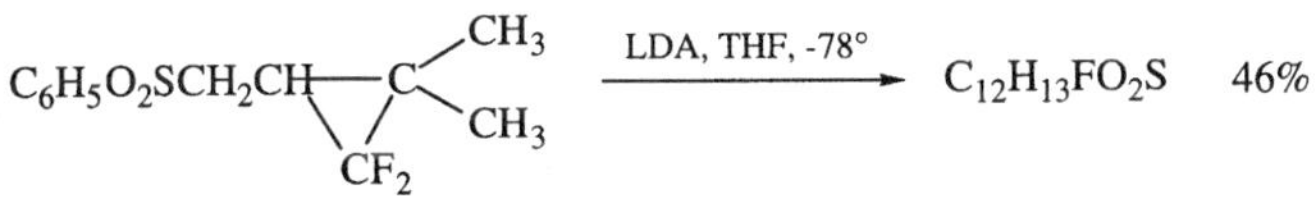

C6H5O2SCH2CH
C
CH3
CH3
CF2
LDA, THF, -78°
C12H13FO2S
46%

SURPRISE 97

Reaction of Diethyl *threo*-2-Bromo-3-fluorosuccinate with Potassium Acetate

What is the product of treatment of diethyl *threo*-α-bromo-α'-fluorosuccinate with potassium acetate?

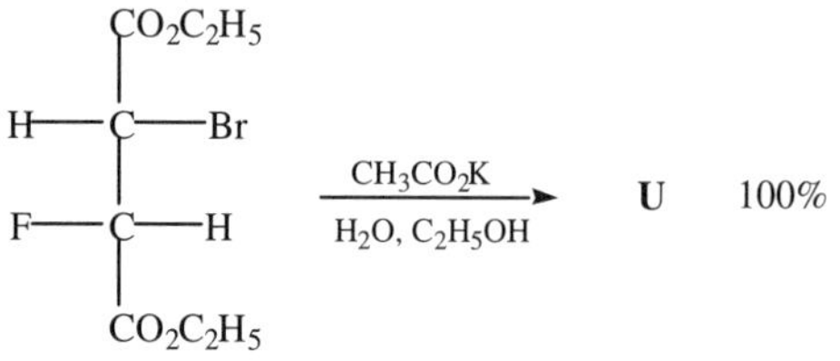

SURPRISE 98

Reaction of 1-Bromo-2-fluorocyclohexanes with Bases

cis-1-Bromo-2-fluorocyclohexane heated with bases readily eliminates hydrogen bromide and gives compounds **V** and **W**.

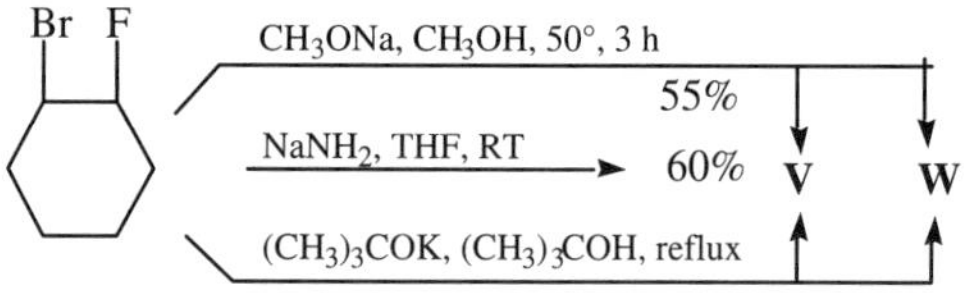

When similar reactions are carried out with *trans*-1-bromo-2-fluoro-cyclohexane, the result of the dehydrohalogenation is quite different, and the products are compounds **W** and **X**. What are these compounds, and how can the difference between the products of elimination of hydrogen halides from either stereoisomer be accounted for?

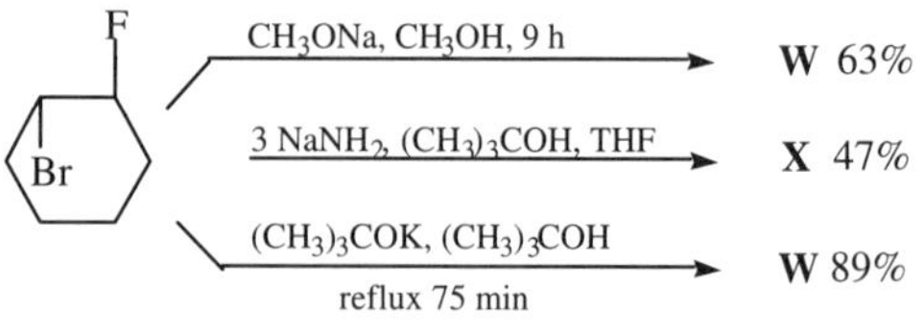

17   Rearrangements

SURPRISE 99

Reaction of Fluorohaloethanes with Aluminum Chloride

What happens when 1-bromo-2-chloro-1,1,2-trifluoroethane is heated with aluminum chloride? What product **Y** is formed?

$$CHClFCBrF_2 \xrightarrow[50°]{AlCl_3} \textbf{Y} \qquad 95\%$$

SURPRISE 100

Treatment of 3-Chloroperfluoropropene with Antimony Halides

What happens when 3-chloroperfluoropropene is heated with antimony trifluoride and chlorine? What is the product **Z**?

$$CF_2{=}CFCClF_2 \xrightarrow[125°]{SbF_3,\ Cl_2} \textbf{Z} \qquad 82\%$$

SURPRISE 101

Reaction of Trifluoronitrosomethane with Ammonia

What happens if trifluoronitrosomethane is treated with ammonia? What is product **A** and what may be a mechanism of its formation?

$$2\ CF_3NO + NH_3 \xrightarrow[3\text{-}4\ h]{RT} \textbf{A} \quad 70\%$$

SURPRISE 102

The Hofmann Degradation of Perfluorobutyramide

Consecutive treatment of perfluorobutyramide with silver oxide, the product **B** with bromine, and the second product **C** with sodium hydroxide gives two different compounds, **D** and **E**, depending on the reaction conditions. What are all four compounds in this reaction?

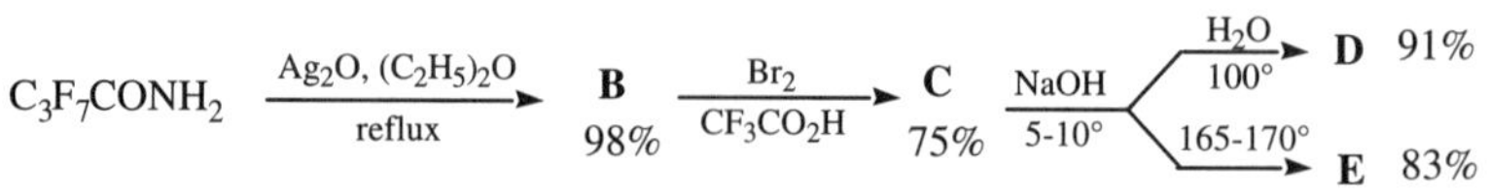

$$C_3F_7CONH_2 \xrightarrow[\text{reflux}]{Ag_2O, (C_2H_5)_2O} \textbf{B} \xrightarrow[\text{CF}_3\text{CO}_2\text{H}]{Br_2} \textbf{C} \xrightarrow[\text{5-10°}]{NaOH}$$

$$\nearrow \xrightarrow[\text{100°}]{H_2O} \textbf{D} \quad 91\%$$

$$\searrow \xrightarrow{165\text{-}170°} \textbf{E} \quad 83\%$$

98%   75%

SURPRISE 103

Irradiation of hexakis(Trifluoromethyl)benzene

Irradiation of hexakis(trifluoromethyl)benzene with ultraviolet light gives three products, **F**, **G**, and **H**. What are their structures?

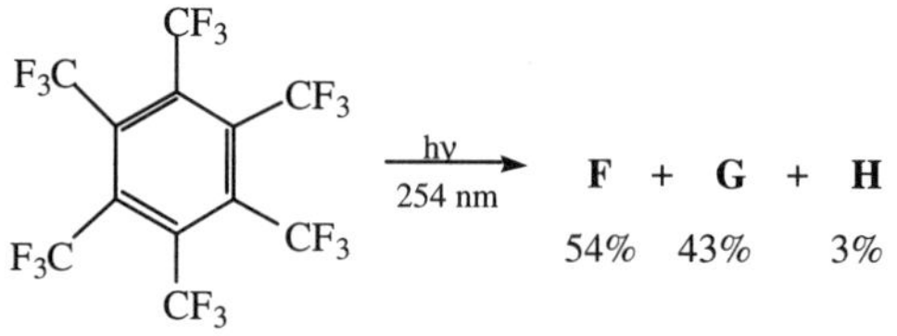

$$\xrightarrow[\text{254 nm}]{h\nu} \quad \textbf{F} + \textbf{G} + \textbf{H}$$

$$54\% \quad 43\% \quad 3\%$$

SURPRISE 104

Reaction of 2-Bromoperfluoronaphthalene with Antimony Pentafluoride

What happens when 2-bromoperfluoronaphthalene is heated with antimony pentafluoride? How is product **I** formed?

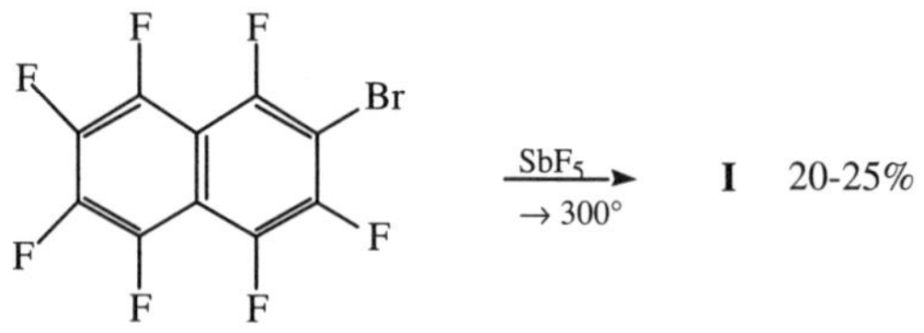

$$\xrightarrow[\rightarrow 300°]{SbF_5} \quad \textbf{I} \quad 20\text{-}25\%$$

Claisen Rearrangement of Fluoroaromatic Acetylenic Ethers

What is the main product **J** of the heating of pentafluorophenyl propargyl ether over silica gell at 370°C?

Part II
Solutions

1   Warmup for Inorganic Chemists

 EXPLANATION 1

Over long periods of time (20–24 years), a slow reaction takes place between anhydrous hydrogen fluoride and the steel of the cylinder. In this way, hydrogen is generated and is responsible for the high pressure inside the cylinder.

$$2\,HF\ +\ Fe\ \longrightarrow\ FeF_2\ +\ H_2$$

Any impending disaster due to this high pressure can be easily averted by bleeding cylinders containing anhydrous hydrogen fluoride periodically after one year, and venting the hydrogen that causes the high pressure. After that, only small amounts of hydrogen fluoride vapors escape from the cylinder when opened at room temperature.

These recent revelations concerning the storage of anhydrous hydrogen fluoride resulted in alerts being sent to companies manufacturing and delivering anhydrous hydrogen fluoride. An alert issued by the Allied Signal Company on 3 February 1997 is reprinted here verbatim:

> Over the years there now have been several incidents reported where very high pressures have been found in steel HF cylinders. In Australia in 1990 an HF cylinder ruptured due to overpressure. Adjacent cylinders which had been filled 20–24 years earlier were found to have internal pressures of over 150 bar (2200 psig). In the UK, some two year old AHF cylinders were found to have internal pressures of over 10 bar (145 psig), and there have been reports of "dormant cylinders exploding" due to this pressure buildup. It has been suggested that the pressure buildup is due to a slow and constant formation over time of hydrogen due to progressive reaction between the HF and the steel cylinder.
>
> The DuPont facility at Parkersburg, West Virginia, recently reported finding very high pressure (~2400 psig) in a small DOT 3E cylinder which had been in storage for perhaps 14 years. The cylinder was safely depressed and the gas vented off was found to be mostly hydrogen. Other similar cylinders were also safely disposed of.

An alert concerning this issue will be circulated to all member companies and will be published in an upcoming issue of *Chemical & Engineering News*.

⇨ EXPLANATION 2

Because the bond dissociation energy of fluorine was not known at that time, its value was estimated by extrapolation of the values for the other three halogens: (63.5 kcal/mol [*1*], or 63–70 kcal/mol [*2*]. It was believed that the fluorine atoms in the molecule of fluorine are held tighter than the other halogen atoms.

Therefore, there was much surprise when the measurements of the bond dissociation energy for fluorine gave a much lower value of 37.5 kcal/mol (157 kJ) [*3*].

Similarly, the original bond dissociation energy of hydrogen fluoride (147.6 kcal/mol (618 kJ) [*1*]) was later corrected to 136 kcal/mol (570 kJ) [*3*].

2 Introduction of Fluorine

⇨ EXPLANATION 3

The mechanism of the unique rearrangement peculiar to fluorination only [*4*] is not unambiguous. Because free acetoxyl radicals are present in the reactions involving lead tetraacetate, and because in the reaction of lead tetraacetate with hydrogen fluoride free fluorine atoms may be anticipated as proven by finding a dimer $(C_6H_5)_2CHFCH_2CH_2$-$CHF(C_6H_5)_2$ in the fluorination of 1,1-diphenylethylene with lead tetrafluoride (generated from anhydrous hydrogen fluoride and lead tetraacetate at –40°C), the intermediate free radical $(C_6H_5)_2CFĊH_2$ can easily rearrange to a more stable free radical $C_6H_5ĊFCH_2C_6H_5$ that then reacts with a fluorine atom to form the final product $C_6H_5CF_2CH_2C_6H_5$ (**C**). The free-radical route is supported by the fact that in the electrolysis of 1,1-diphenylethylene in anhydrous hydrogen fluoride, where fluorine atoms are present, the rearranged product 1,1-difluoro-1,2-diphenyl was also found [*5*].

$(C_6H_5)_2C{=}CH_2 \xrightarrow[0°]{F} (C_6H_5)_2CF\overset{\bullet}{C}H_2 \longrightarrow C_6H_5\overset{\bullet}{C}FCH_2C_6H_5 \xrightarrow{F} C_6H_5CF_2CH_2C_6H_5$

$F_2, -40°$

$(C_6H_5)_2CF\overset{\bullet}{C}H_2 \longrightarrow (C_6H_5)_2CFCH_2CH_2CF(C_6H_5)_2$

On the other hand, explanation of the reaction of 1,1-diphenylethylene with aryl iodide difluoride by an ionic mechanism has also some merit [6]. It is quite possible that the two reactions proceed to the same product by two different mechanisms.

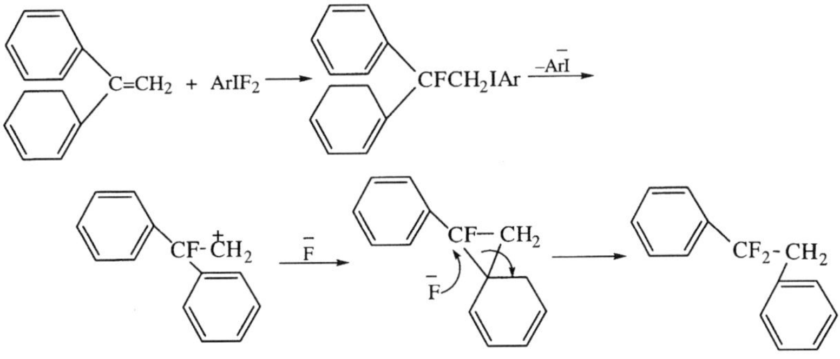

A = $(C_6H_5)_2CFCH_2F$ **B** = $(C_6H_5)_2C=CHF$ **C** = $C_6H_5CF_2CH_2C_6H_5$

A rearrangement similar to that described in *Surprise 3* occurs when 1,1-bis(*p*-chlorophenyl)-2,2,2-trichloroethane (DDT) is treated with anhydrous hydrogen fluoride and red mercuric oxide [7]. Intermediates assumed in this reaction are shown in the following equations.

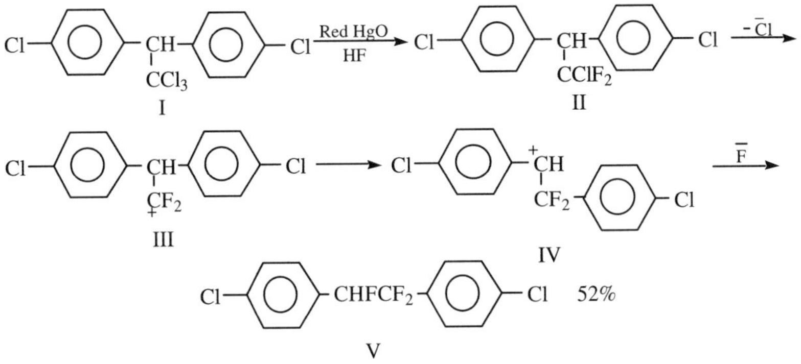

The intermediate **II** is assumed to dissociate and release chlorine anion, leaving cation **III** that rearranges prior to its combination with the fluoride anion present in the reaction mixture (**IV** and **V**) [7]. Similar reactions are involved in the conversion of 1,1-bis(*p*-chlorophenyl)-1,2,2,2-tetrachloroethane.

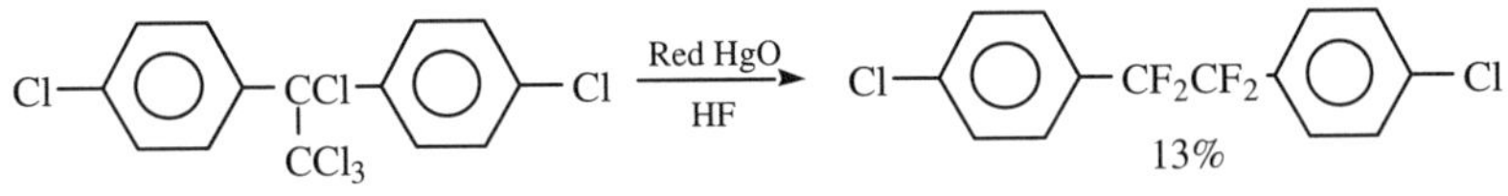

The rearrangement encountered in the fluorination of 1,1-diphenyleth-ylene with lead tetrafluoride or aryl iodide difluoride does not occur when other halogens are used. However, 1,1-bis(p-chlorophenyl)-2,2-dichloro-ethane (VI) and 1,1-bis(p-chlorophenyl)-1,2,2,2-tetrachloroethane (VII) rearrange to 1,2-bis(p-chlorophenyl)-1,2-dichloroethane (VIII) and 1,2-bis(p-chlorophenyl)-1,1,2,2-tetrachloroethane (IX), respectively, when treated with ferric chloride [7].

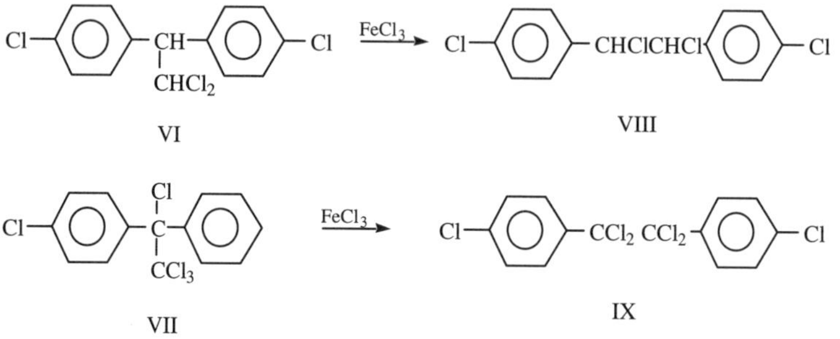

EXPLANATION 4

1. Diethylaminosulfur trifluoride reacts with the 2-hydroxyl to form an intermediate (V) in which the neighboring benzyloxy group participates, forming a three-membered epoxide ring (VI). This is cleaved by the fluo-ride ion detached from the intermediate VI, and the result is a product with retention of configuration, benzyl 3-azido-4,6-O-benzylidene-2,3-dideoxy-2-fluoro-α-D-altropyranoside II.

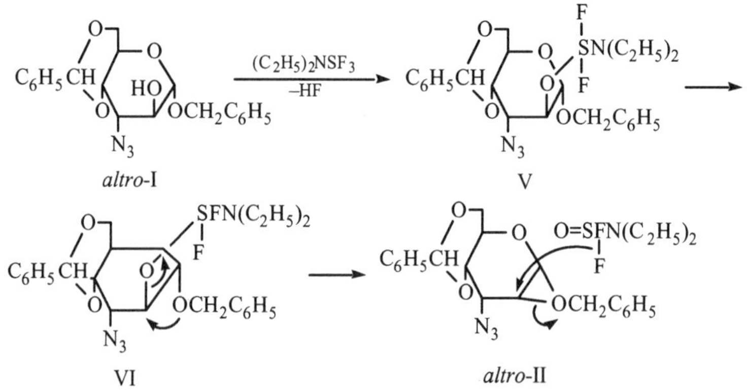

2. The intermediate VI formed from the starting material and diethyl-aminosulfur trifluoride undergoes an S_Ni reaction in an eight-membered transition state involving the azido group. The replacement occurs with

inversion of configuration on carbons 2 and 3, and gives benzyl 2-azido-4,6-O-benzylidene-2,3-dideoxy-3-fluoro-α-D-glucopyranoside III.

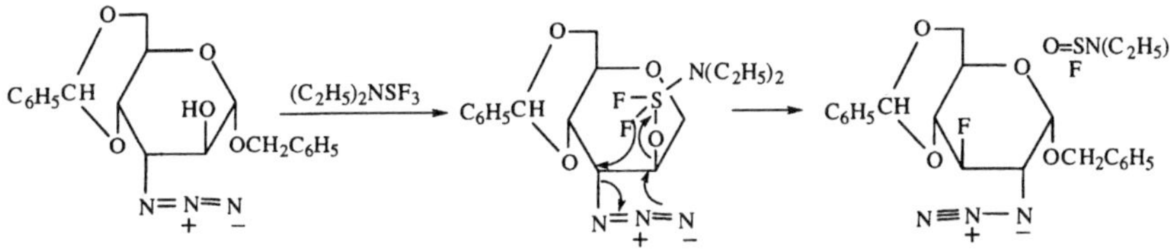

C_6H_5CH ... $\xrightarrow{(C_2H_5)_2NSF_3}$ ... C_6H_5CH ... $\longrightarrow$ ... C_6H_5CH

allo-I   VI   *gluco*-III

The intermediate VII formed from the starting compound and diethyl-aminosulfur trifluoride reacts by an S_N2 mechanism to yield a product with inversion of configuration on carbons 1 and 2, 2-benzyl-3-azido-4,6-O-benzylidene-3-deoxy-β-D-allopyranosyl fluoride (IV) [8].

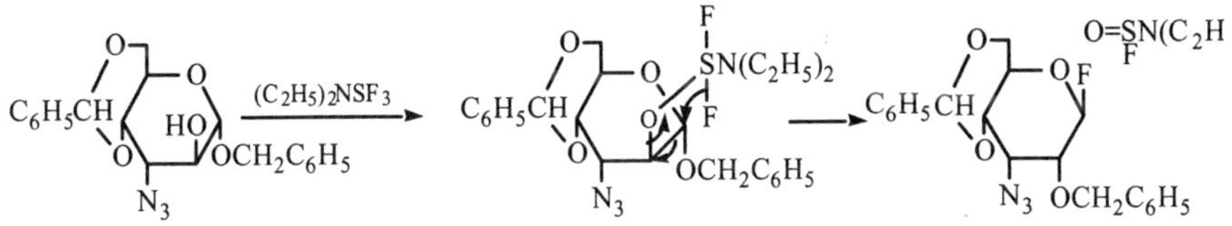

C_6H_5CH ... $\xrightarrow{(C_2H_5)_2NSF_3}$ ... C_6H_5CH ... $\longrightarrow$ ... C_6H_5CH

altro-I   VII   *allo*-IV

⟴ EXPLANATION 5

Ethyl acetoacetate reacts in its enol form with one molecule of diethyl-aminosulfur trifluoride. The hydroxyl group is replaced by fluorine, and ethyl α-fluoroacetoacetate is formed. This compound reacts again in its enol form with another molecule of diethylaminosulfur trifluoride to give ethyl α,β-difluorocrotonate as a mixture of equal parts of E- and Z-isomer.

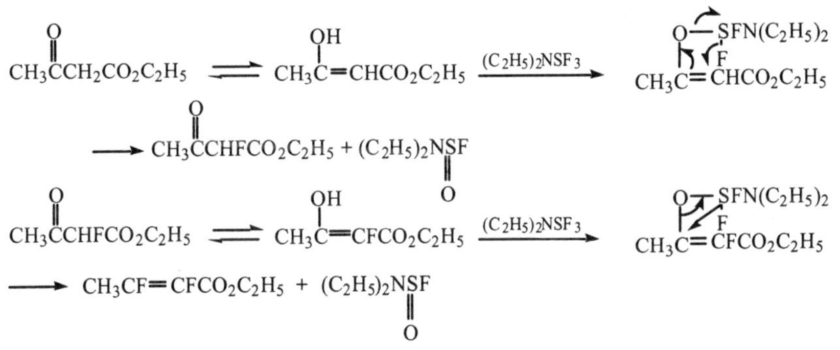

$$CH_3CCH_2CO_2C_2H_5 \rightleftharpoons CH_3C=CHCO_2C_2H_5 \xrightarrow{(C_2H_5)_2NSF_3} CH_3C=CHCO_2C_2H_5$$

$$\longrightarrow CH_3CCHFCO_2C_2H_5 + (C_2H_5)_2NSF \qquad [9]$$

$$CH_3CCHFCO_2C_2H_5 \rightleftharpoons CH_3C=CFCO_2C_2H_5 \xrightarrow{(C_2H_5)_2NSF_3} CH_3C=CFCO_2C_2H_5$$

$$\longrightarrow CH_3CF=CFCO_2C_2H_5 + (C_2H_5)_2NSF$$

Similar reactions of diethylaminosulfur trifluoride take place with 2,4-pentanedione (acetylacetone), with butyl acetoacetate, and with ethyl 3-oxooctanoate, giving the corresponding difluoro compounds as *E/Z* mixtures in the ratio of 1:1 [9].

$$CH_3COCH_2COCH_3 \xrightarrow{(C_2H_5)_2NSF_3} E/Z\ CH_3CF{=}CFCOCH_3$$

$$CH_3COCH_2CO_2C_4H_9 \xrightarrow{(C_2H_5)_2NSF_3} CH_3CF{=}CFCO_2C_4H_9$$

$$CH_3(CH_2)_4COCH_2CO_2C_2H_5 \xrightarrow{(C_2H_5)_2NSF_3} E/Z\ CH_3(CH_2)_4CF{=}CFCO_2C_2H_5$$

3   Reductions

⟢ EXPLANATION 6

Addition of hydrogen to double bonds in catalytic hydrogenation is very easy, certainly much easier than hydrogenolysis of carbon-halogen bonds. Whereas the carbon-halogen bond in saturated fluoro- and chloro compounds is hydrogenolyzed only under forcing conditions, hydrogenolysis of unsaturated fluoro compounds occurs very easily. Hydrogenolysis of alkyl fluorides (methyl, ethyl, propyl, and isopropyl fluorides) requires temperatures of 110–240°C [10]. In contrast, hydrogenolysis of vinyl fluorides, benzyl fluorides, and aromatic fluorides occurs very readily at room temperature and atmospheric pressure. Naturally double bonds are hydrogenated under these conditions [5, 11, 12, 13].

$$CF_3CF{=}CHF \xrightarrow[20°]{H_2/Pd(Al_2O_3)} \underset{60\%}{CF_3CHFCH_2F} + \underset{30\%}{CF_3CHFCH_3}$$

[12]

$$(CF_3)_2C{=}CF_2 \xrightarrow[100°]{H_2/Ni} \underset{10\%}{(CF_3)_2CHCHF_2} + \underset{75\%}{(CF_3)_2CHCH_3}$$

The compound **D** resulting from catalytic hydrogenation of *trans*-α-fluorostilbene is bibenzyl [5], and compounds **E** and **F** resulting from hydrogenation of difluoromaleic acid are fluorosuccinic acid and succinic acid, respectively [11]. Naturally similar hydrogenolyses of vinylic fluorine take place in the hydrogenation of difluorofumaric acid, fluoromaleic and fluorofumaric acid, and many other unsaturated fluorides [11,12,13].

D $C_6H_5CH_2CH_2C_6H_5$ **E** $HO_2CCH_2CHFCO_2H$ **F** $HO_2CCH_2CH_2CO_2H$

The facile replacement of vinylic, benzylic, and aromatic fluorines [11] can be rationalized by a concept of simultaneous (conjugate) reaction involving participation of π-bonds in five- and six-membered transition states [11,13].

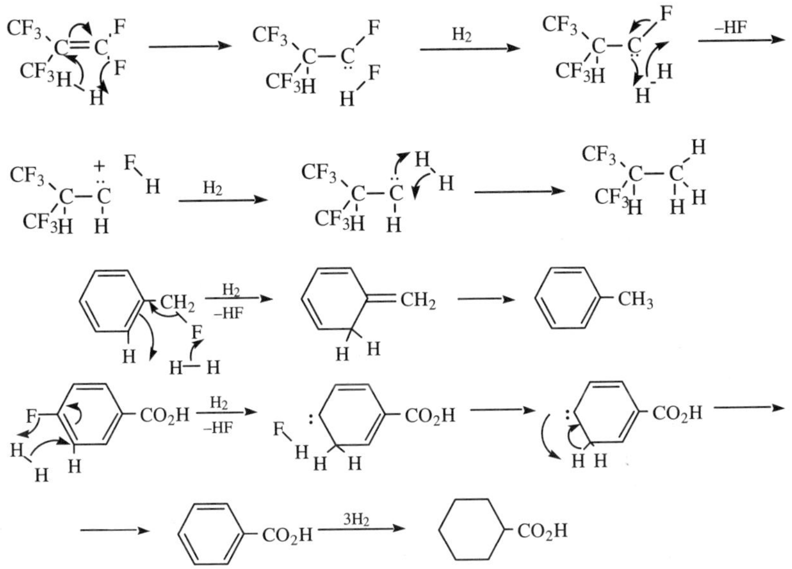

➭ **EXPLANATION 7**

In catalytic hydrogenation, chlorine is replaced by hydrogen much more readily than fluorine because the carbon-chlorine bond breaks more easily than the carbon-fluorine bond. Therefore, the product **G** is 2,4,5,6-tetrafluoropyridine [14].

In the reduction with lithium aluminum hydride (and other complex hydrides), replacement of a halogen occurs by a nucleophilic attack of the reagent's anion, $\overline{A}lH_4$. This anion joins that carbon of the pyridine ring that is more electrophilic, that is, the carbon linked to fluorine. Fluorine, because of its very strong electronegativity and back-donation of electrons, polarizes the carbon-halogen bond more strongly than does chlorine, and the hydride anion combines in a rate-determining step with the carbon holding fluorine. In the formation of such a transition state, the aromaticity of the pyridine is destroyed and is restored by ejection of fluoride ion in a fast reaction. As a consequence, compound **H** is 3-chloro-2,5,6-trifluoropyridine [15].

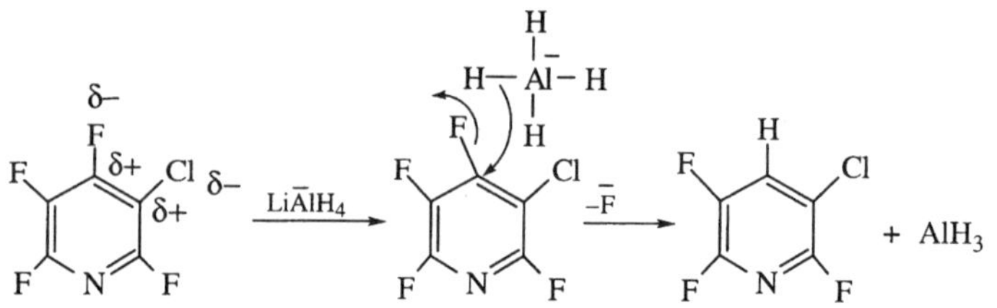

G [14]

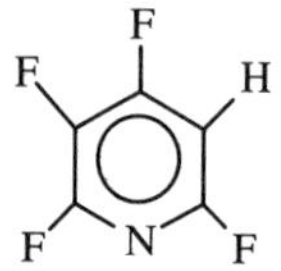

H [15]

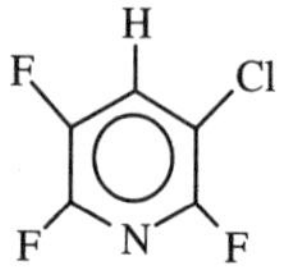

EXPLANATION 8

Irradiation of alcohols with ultraviolet light of 300 or 254 nm initiates a free-radical reaction resulting in the formation of a ketyl radical, formed by abstraction of hydrogen from the carbon atom linked to the oxygen. The alcohol best suited for this purpose is isopropyl alcohol, which transfers the hydrogen from its oxygen to the reducible substrate, and is converted to acetone. Under such conditions, isopropyl alcohol reduces halogen compounds including fluoro compounds, but in compounds containing fluorine and other halogens, chlorine, bromine, and iodine are replaced by hydrogen in preference to fluorine.

Compound **I** in the above reaction is cyclohexanone [16], and compound **J** is methyl 2,3-dichloro-3,3-difluoropropionate [17].

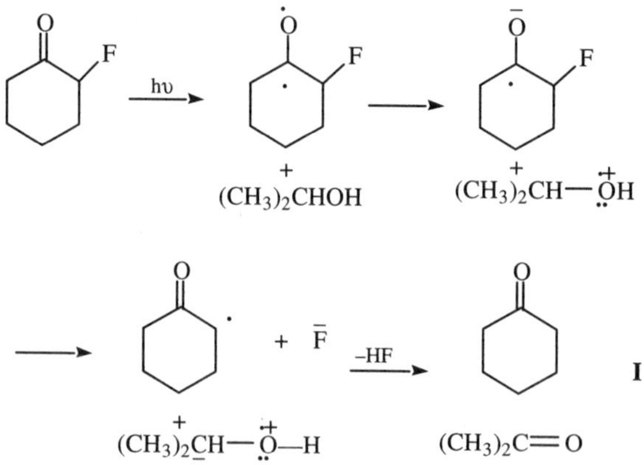

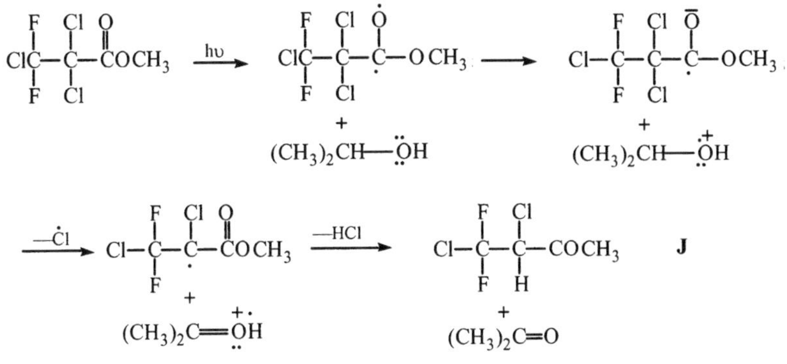

$$\underset{\underset{F}{\underset{|}{\overset{|}{\overset{F}{\mid}}}}{\overset{Cl}{\underset{|}{\overset{Cl}{\mid}}}}{\text{ClC–C–COCH}_3} \xrightarrow{h\upsilon} \underset{\underset{F}{|}}{\overset{F}{|}}\underset{\overset{Cl}{|}}{}\text{ClC–C–}\overset{\cdot}{\text{C}}\text{–OCH}_3 \longrightarrow \text{Cl–C–C–}\overset{\bar{\text{O}}}{\text{C}}\text{–OCH}_3$$

$$+$$
$$\text{(CH}_3)_2\text{CH}\!—\!\overset{\cdot\cdot}{\underset{\cdot\cdot}{\text{O}}}\text{H} \qquad\qquad \text{(CH}_3)_2\text{CH}\!—\!\overset{+}{\underset{\cdot\cdot}{\text{O}}}\text{H}$$

$$\xrightarrow{-\overset{\cdot}{\text{Cl}}} \text{Cl–C–C–COCH}_3 \xrightarrow{-\text{HCl}} \text{Cl–C–C–COCH}_3 \quad \textbf{J}$$

$$+$$
$$\text{(CH}_3)_2\text{C}\!=\!\overset{+}{\underset{\cdot\cdot}{\text{O}}}\text{H} \qquad\qquad \text{(CH}_3)_2\text{C}\!=\!\text{O}$$

4   Oxidations

⇨ EXPLANATION 9

The primary oxidation product, chlorotrifluoroethylene oxide (compound **K**), can isomerize either to trifluoroacetyl chloride, or to chlorodifluoroacetyl fluoride (compound **L**). It is unfortunate that the isomerization occurs in this way, because the other possibility would lead to a very coveted compound, trifluoroacetic acid [18, 19, 20].

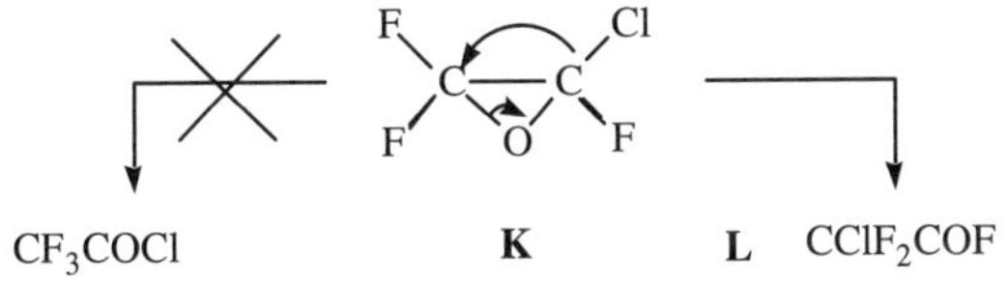

$$\text{CF}_3\text{COCl} \qquad\qquad \textbf{K} \qquad \textbf{L} \quad \text{CClF}_2\text{COF}$$

[20]

⇨ EXPLANATION 10

Lead tetraacetate cleaves the carbon-carbon bonds between two vicinal functions (hydroxyls or carbonyls). It is very useful in saccharide chemistry.

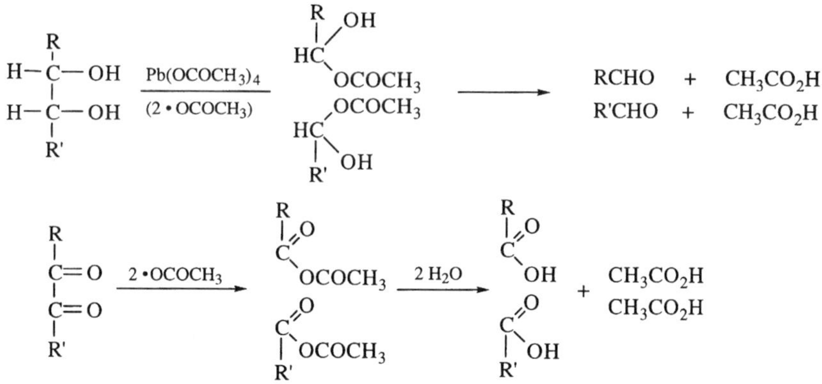

The reaction schemes showing the mechanism (top of page):

$$H-\underset{\underset{R'}{|}}{\overset{\overset{R}{|}}{C}}-OH \quad\underset{\underset{R'}{|}}{\overset{\overset{R}{|}}{C}}-OH \quad\xrightarrow[(2 \cdot OCOCH_3)]{Pb(OCOCH_3)_4}\quad \overset{R}{\underset{R'}{HC}}\overset{OH}{\underset{OH}{}}\begin{matrix}OCOCH_3\\OCOCH_3\end{matrix} \quad\longrightarrow\quad \begin{matrix}RCHO & + & CH_3CO_2H\\ R'CHO & + & CH_3CO_2H\end{matrix}$$

$$\underset{\underset{R'}{|}}{\overset{\overset{R}{|}}{\underset{C=O}{C=O}}}\quad\xrightarrow{2\cdot OCOCH_3}\quad \begin{matrix}R & \\ C \diagdown O \\ OCOCH_3 \\ C \diagup O \\ OCOCH_3 \\ R'\end{matrix}\quad\xrightarrow{2\,H_2O}\quad \begin{matrix}R\\ C \diagdown O\\ OH\\ C\diagup O\\ OH\\ R'\end{matrix}\quad+\quad \begin{matrix}CH_3CO_2H\\ CH_3CO_2H\end{matrix}$$

In the compound described above, two acetoxyl radicals are probably added to the double bond between the two carbons linked to the two amino groups. Ejection of two molecules of acetic acid generates a diimino compound which reacts with other two acetoxyl radicals to cleave the ring and give a straight-chain dinitrile [21].

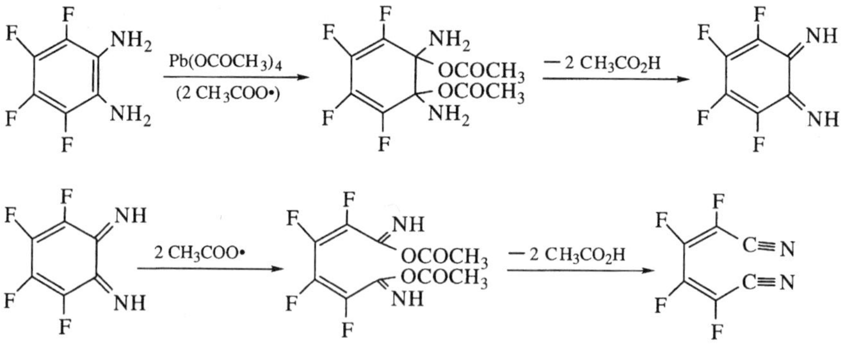

First reaction row of fluorinated ring structures:

(ring with F substituents and two NH₂ groups) $\xrightarrow[(2\,CH_3COO\cdot)]{Pb(OCOCH_3)_4}$ (ring with NH₂, OCOCH₃, OCOCH₃, NH₂) $\xrightarrow{-2\,CH_3CO_2H}$ (ring with two NH groups)

Second reaction row:

(ring with two NH groups) $\xrightarrow{2\,CH_3COO\cdot}$ (ring with NH, OCOCH₃, OCOCH₃, NH) $\xrightarrow{-2\,CH_3CO_2H}$ (ring with two $C\equiv N$ groups)

EXPLANATION 11

First the diamino compound is diazotized to give a bis-diazonium compound.

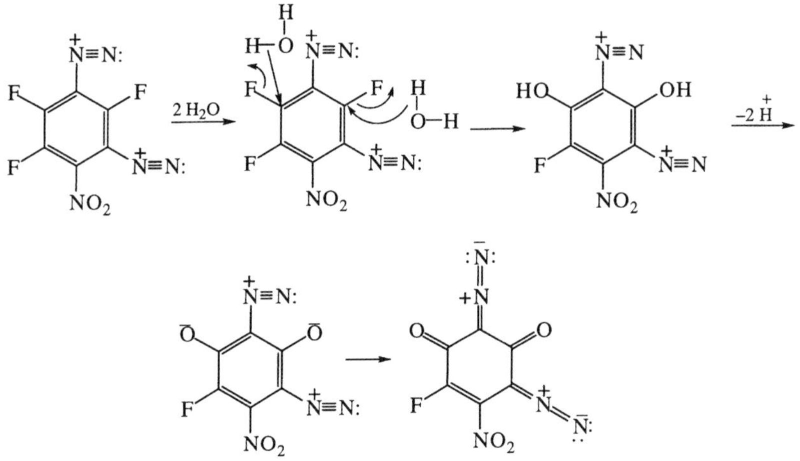

The strongly electronegative diazonium groups activate (like nitro groups do) the vicinal positions in the benzene ring for nucleophilic reaction with water. Thus, fluorines in positions 2 and 6 are replaced by hydroxyls. Ejection of phenolic hydrogens converts the diazonium groups to diazo groups, and the compound to a diazooxide [22].

5 Preparation of Halogen Derivatives

⟶ EXPLANATION 12

The bond in iodine fluoride, like in any other halogen fluoride, is polarized in such a way that iodine (or bromine or chlorine) is partially positive and fluorine partially negative. The double bond in chlorotrifluoroethylene is also polarized. Back-donation of electrons of fluorine to the double bond makes the carbon holding two fluorines more electrophilic than the other carbon, which acquires a partial negative charge. It is to be expected that the positively charged iodine would join the carbon carrying chlorine and fluorine, and the fluoride anion will join the carbon with two fluorines. Such addition does occur, but is accompanied by the predominant addition of iodine to the carbon carrying two fluorines, and the fluoride anion to the opposite carbon. Bromine fluoride generated from bromine trifluoride and bromine behaves similarly [23].

There is no explanation of this "reverse" addition of halogen fluorides. It is possible that the carbon holding chlorine is sterically less accessible to the voluminous iodine or bromine. Another explanation for the formation of two products is the common justification that the reaction is not regiospecific. In contrast, addition of iodine fluoride to perfluoropropene takes place as expected.

$$I-F \longrightarrow \overset{\delta+}{I}-\overset{\delta-}{F} \longrightarrow \overset{+}{I}-\overset{-}{F} \qquad CClF=CF_2 \longrightarrow \overset{\delta-}{C}ClF=\overset{\delta+}{C}F_2$$

$$\xrightarrow[\;20°\;]{BrF_3,\ Br_2} \quad \overset{\delta-}{C}ClF=\overset{\delta+}{C}F_2 \quad \xrightarrow[\;20°\;]{IF_5,\ 2\ I_2}$$

$$CBrClFCF_3 \ + \ CClF_2CF_2Br \qquad\qquad CClFICF_3 \ + \ CClF_2CF_2I$$

$$\quad\; 13\% \qquad\qquad\quad 73\% \qquad\qquad\qquad\quad 37\% \qquad\qquad\quad 45\%$$
$$\qquad\qquad\qquad\qquad\qquad\qquad\qquad\qquad\qquad \textbf{M} \qquad\qquad\qquad \textbf{N}$$

$$CF_3CF=CF_2 \longleftrightarrow CF_3\overset{\delta-}{C}F-\overset{\delta+}{C}F_2 \xrightarrow{\ IF\ } CF_3CFICF_3 \quad 99\%$$

⟵ EXPLANATION 13

The double bond in chlorotrifluoroethylene is polarized in such a way that the carbon holding two fluorines is less electronegative (more electrophilic) than the carbon with chlorine and fluorine. Sounds absurd? Not really, because the back-donation of electrons of fluorine makes the bond more electronegative at the other end. Under polar conditions, proton, the attacking species, joins the partially electronegative carbon holding chlorine and fluorine, and the bromide anion joins the carbon with two fluorines [24, 25, 26, 27].

In the case of a free-radical addition of hydrogen bromide, a type of addition limited to hydrogen bromide only, the attacking species is not proton, but a bromine atom that will join the carbon in such a way as to generate a more stable free-radical species. It will add to the carbon carrying two fluorines. In the intermediate free radical, the single electron is better accommodated at the carbon holding chlorine, because chlorine can disperse the electrons in its d-orbitals.

The intermediate free radical abstracts hydrogen atom from a molecule of hydrogen bromide, thus generating atomic bromine, which perpetuates the chain reaction. [24]

$$\cdot CClF-CBrF_2 \xrightarrow{HBr} CHClF-CBrF_2 + \cdot Br$$

$$CClF=CF_2 \xrightarrow{\cdot Br} \cdot CClF-CBrF_2 \xrightarrow{HBr} CHClF-CBrF_2 + \cdot Br$$

 EXPLANATION 14

Trifluoromethyl group in trifluoropropene affects the double bond in two ways. The overall effect is drainage of electrons from the double bond so that the electron density is lowered, and consequently electrophilic reactions such as additions of hydrogen halides are difficult to achieve. Moreover, the electron density at the carbon adjacent to the trifluoromethyl group is higher relative to that at the other end of the double bond. Since addition of hydrogen halides starts with a combination of proton with the more electronegative pole of the double bond, proton joins the carbon adjacent to the trifluoromethyl cluster, and the chloride anion is linked to the more distant carbon. Thus, the addition occurs differently from that of the parent hydrocarbon, and is sometimes referred to as the "anti-Markovnikov" addition. It requires a strong catalysis by aluminum chloride. Other catalysts such as zinc chloride, boron trifluoride, bismuth trichloride, and ferric chloride are ineffective [28]. Hydrogen bromide under catalysis with aluminum bromide adds in the same direction.

$$CH_2=CHCF_3 \longleftrightarrow \overset{+}{C}H_2-\bar{C}HCF_3$$

$$\xrightarrow[100°,\,9\,h]{HCl,\,AlCl_3} CH_2ClCH_2CF_3 \quad 33\%$$

$$\xrightarrow[100°,\,9\,h]{HBr,\,AlBr_3} CH_2BrCH_2CF_3 \quad 69\%$$

 EXPLANATION 15

Addition of hydrogen halides across the double bond in perfluoropropene resembles the addition to 3,3,3-trifluoropropene. Here, too, fluorines drain electrons from the double bond, rendering it electron-poor, which means that electrophilic additions of hydrogen halides are very difficult. Electron density of the double bond is distributed in such a way that the partial negative charge is at the carbon adjacent to the trifluoromethyl group. Consequently, the proton of the hydrogen halides joins this carbon while the halide anion adds to the terminal end. Again, the addition could be called "anti-Markovnikov" [29].

$$\underset{F}{\overset{F}{\diagdown}}C=CFCF_3 \longleftrightarrow :\overset{\cdot\cdot+}{F}\underset{F}{\overset{}{\diagdown}}C-\overset{-}{C}FCF_3 \longleftrightarrow \overset{\delta+ \quad \delta-}{CF_2-CF-CF_3}$$

$$CF_2=CFCF_3$$

$\xrightarrow[\text{200}°]{\text{HF, CaSO}_4\text{, C}}$ CF$_3$CHFCF$_3$ 80%

$\xrightarrow[\text{230}°]{\text{HCl}}$ CClF$_2$CHFCF$_3$ 60%

$\xrightarrow{\text{HBr}}$ CBrF$_2$CHFCF$_3$ 50%

If the addition of hydrogen bromide to perfluoropropene is carried out under the conditions favoring a free-radical mechanism, both possible products are formed [30].

$$CF_2=CFCF_3 \xrightarrow[\text{UV or }\gamma\text{-rays}]{\text{HBr}} CBrF_2CHFCF_3 + CHF_2CHBrCF_3$$
$$\qquad\qquad\qquad\qquad\qquad\quad 30\% \qquad\qquad 22\%$$

➦ EXPLANATION 16

Double bonds in perfluoroalkenes are very electrophilic because their electron density is low owing to the combined inductive and resonance effects of fluorine atoms.

$$\underset{F}{\overset{:\overset{\cdot\cdot}{F}}{\diagdown}}C=CF-CF_3 \longleftrightarrow \underset{F}{\overset{\overset{+}{F}}{\diagdown}}C-\overset{-}{C}F-CF_3$$

For this reason, electrophilic additions such as addition of hydrogen halides are very difficult, as can be seen in Explanations 13, 14, and 15. But because of the low overall electron density, the double bonds in perfluoroalkenes are easily attacked by nucleophiles such as fluoride ion. If perfluoropropene is treated with dry potassium fluoride in a proton-donor solvent such as formamide, the fluoride anion joins the less negative carbon of the double bond and forms a negatively charged intermediate that readily reacts with the proton of formamide to give 2 H-perfluoropropane [31]. Such a "nucleophilic" addition takes place with other fluorides such as cesium fluoride, tetraethylammonium fluoride, and silver fluoride. In the last case, the addition of hydrogen fluoride can be achieved in two separate steps. Under anhydrous conditions, silver fluoride adds to the double bond and creates an organometallic compound, which is then decomposed with water to yield 2 H-perfluoropropane [31, 32, 33].

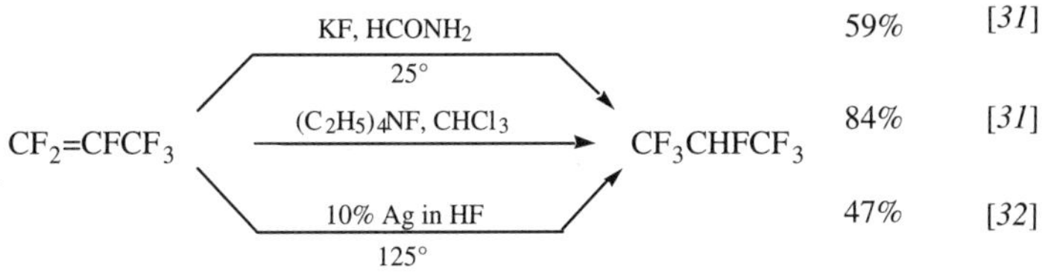

$$CF_2=CFCF_3 \xrightarrow[25°]{AgF, CH_3CN} CF_3CFAgCF_3 \xrightarrow{H_2O} CF_3CHFCF_3 + AgOH$$
$$95\%$$

➯ EXPLANATION 17

Despite the general belief that halogens deactivate aromatic rings in electrophilic substitutions, the major product of chlorination of *o*-fluorotoluene is 5-chloro-2-fluorotoluene, the result of replacement of the hydrogen *para* to the fluorine (**O**), not *para* to the methyl group. Only a smaller fraction results from chlorination *ortho* to the methyl group (**P**).

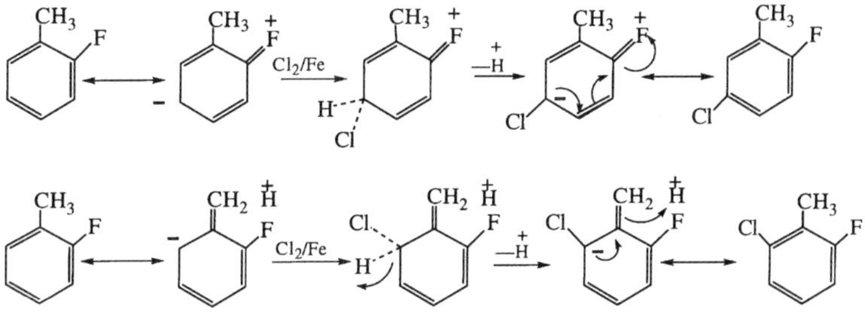

In electrophilic aromatic substitutions, alkyl groups increase the electron density of the aromatic ring and thus activate the ring for electrophilic attack. Halogens by their inductive effect deactivate the rings. The exception is fluorine, whose strong back-donation of electrons activates the aromatic ring. Both methyl group and fluorine activate *para* positions, and activation by fluorine is stronger than that of the methyl group, as evidenced by the result of the chlorination [34].

In contrast to chlorine, bromine, and iodine, which deactivate aromatic rings for electrophilic substitutions, fluorine exerts an even higher activation of the ring than does hydrogen, as has been shown by measurements of the rates of bromination of durene (1,2,4,5-tetramethylbenzene) and its 3-halo derivatives [35].

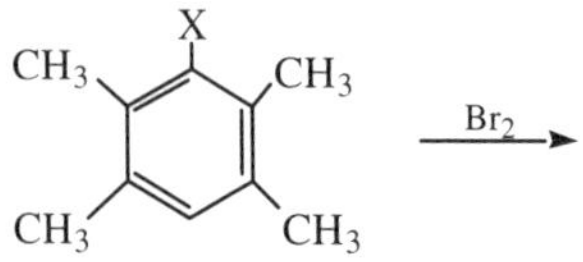

X = H	F	Cl	Br	I	Reaction
1000	2310	72.6	30.9	40.0	rates

⇨ EXPLANATION 18

At 400°C and without a catalyst, chlorine reacts in its atomic form and causes a free-radical substitution. Of all the bonds to carbons of the benzene ring, the bond between carbon and nitrogen is the weakest (292 kJ, 70 kcal/mol) compared to carbon-hydrogen bond (415 kJ, 99 kcal/mol) and carbon-fluorine bond (443 kJ, 106 kcal/mol). Therefore, atomic chlorine cleaves the bond between carbon and nitrogen and replaces the nitro group. The product is 1-chloro-3,4-difluorobenzene (compound **Q**) [36].

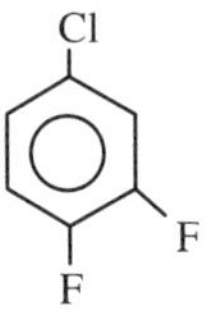

⇨ EXPLANATION 19

Nitro groups in aromatic rings activate substituents in *ortho* and *para* positions for nucleophilic displacements. The nucleophile here is a chloride anion that replaces fluorine. Of the five nitro groups, those in both *ortho* and *para* positions to fluorine activate fluorine for nucleophilic displacement very strongly, so that even a relatively weak nucleophile like chloride replaces fluorine under very mild conditions. Compound **R** is chloropentanitrobenzene [37].

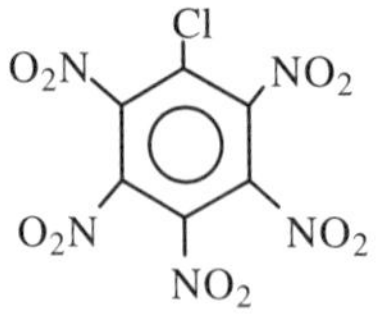

➦ EXPLANATION 20

Trifluoroacetyl hypochlorite dissociates into chlorine cation and trifluoroacetoxy anion:

$$CF_3C \overset{O}{\underset{OCl}{\diagup}} \longrightarrow CF_3C \overset{O}{\underset{O^-}{\diagup}} + \overset{+}{Cl}$$

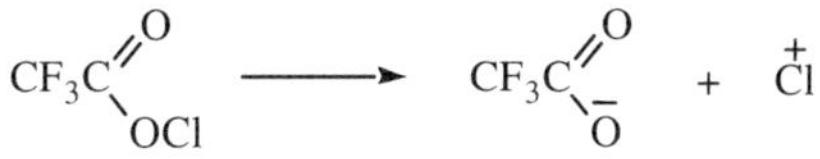

The double bond in 1,1-difluoroethylene is polarized by fluorine in such a way that the carbon carrying two fluorine atoms is less electronegative than the carbon with hydrogen atoms. Consequently, the chlorine cation joins the more electronegative carbon of the double bond and forms a three-membered ring containing a positively charged chlorine. The ring is subsequently cleaved by the trifluoroacetoxy anion at the difluoromethylene group [38].

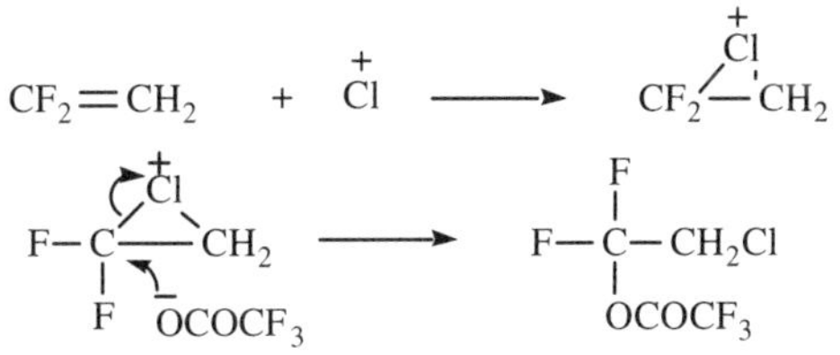

$$CF_2{=}CH_2 \quad + \quad \overset{+}{Cl} \longrightarrow CF_2 \overset{\overset{+}{Cl}}{\diagdown} CH_2$$

$$\underset{F \quad OCOCF_3}{F{-}C{\overset{\overset{+}{Cl}}{\text{—}}}CH_2} \longrightarrow \underset{OCOCF_3}{F{-}\overset{F}{\underset{|}{C}}{-}CH_2Cl}$$

$$C_4H_2ClF_5O_2 \;=\; CF_3COOCF_2CH_2Cl$$

➦ EXPLANATION 21

Reaction of pentafluorophenol with *tert*-butyl hypobromite starts as an electrophilic substitution in the benzene ring. The electrophile is formed by dissociation of *tert*-butyl hypobromite to *tert*-butoxy anion and bromine cation. The bromine cation attacks the *para* position and forms a positively charged "Wheland complex," a nonaromatic species that is converted by ejection of proton from the phenolic hydroxyl to a quinonoid compound, 4-bromopentafluorocyclohexa-2,5-dienone [39].

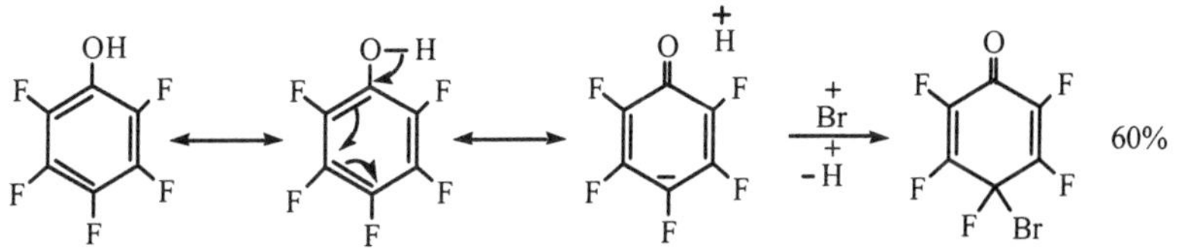

EXPLANATION 22

When hexafluorocyclobutene is treated with aluminum chloride, all fluorine atoms are replaced by chlorine. On treatment with aluminum bromide, all are replaced by bromine. The products are hexachlorocyclobutene and hexabromocyclobutene, respectively. The reason for this quite unexpected reaction may be the difference in the strength of carbon-fluorine bond and aluminum-fluorine bond. As the aluminum-fluorine bond (595 kJ/mol, 142 kcal) is stronger than the carbon-fluorine bond (443 kJ/mol, 106 kcal), the lower reaction enthalpy may be the driving force for the halogen exchange [40].

$$
\mathbf{S} \quad \begin{array}{c} CCl{=}CCl \\ | \quad\quad | \\ CCl_2 \cdot CCl_2 \end{array}
\qquad\qquad
\mathbf{T} \quad \begin{array}{c} CBr{=}CBr \\ | \quad\quad | \\ CBr_2 \cdot CBr_2 \end{array}
$$

Surprisingly, when 1,2-dichlorohexafluorocyclopentene is treated with aluminum bromide, the fluorine atoms are exchanged for bromine, but the chlorine atoms remain intact [40].

$$
\begin{array}{c} CCl{:}CCl \\ / \qquad \backslash \\ CF_2 \quad\quad CF_2 \\ \quad CF_2 \end{array}
\quad \xrightarrow{\text{AlBr}_3} \quad
\begin{array}{c} CCl{:}CCl \\ / \qquad \backslash \\ CBr_2 \quad\quad CBr_2 \\ \quad CBr_2 \end{array}
$$

 EXPLANATION 23

As in Explanation 22, here, too, a halogen exchange takes place. Only the fluorine atoms adjacent to the oxygen are replaced by chlorine. The result is perfluoro-2-butyl-2,5,5-trichlorofuran [41]. The possible reason for the halogen exchange is explained in Surprise 22.

$$
\begin{array}{c} CF_2 {-} CF_2 \\ / \qquad\qquad \backslash \\ CCl_2 \qquad\quad CClC_4F_9 \\ \backslash \quad\;\; / \\ O \end{array}
$$

 EXPLANATION 24

Like in the Surprises 22 and 23, aluminum chloride replaces fluorine by chlorine. In this case, like in Surprise 23, only the fluorine atoms adjacent to the oxygen are exchanged. In contrast to Surprise 23, the ring is cleaved, and the product is 5,5,5-trichlorohexafluoropentanoyl chloride [42].

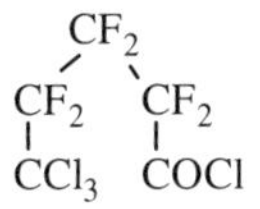

$$\begin{array}{c} CF_2 \\ {}^{\diagup} \quad {}^{\diagdown} \\ CF_2 \quad CF_2 \\ | \qquad | \\ CCl_3 \quad COCl \end{array}$$

6 Nitration

⟴ EXPLANATION 25

In nitration, an electrophilic aromatic substitution, the electrophile is a nitronium cation. It attacks the benzene ring in the place of the highest electron density. In *o*-fluorotoluene, there are several positions of high electron density: *ortho* and *para* with respect to the methyl group, and *ortho* and *para* with respect to fluorine. Both substituents, methyl and fluorine, direct the entering electrophile to the activated positions. Whose influence is stronger, that of the methyl group, or that of fluorine?

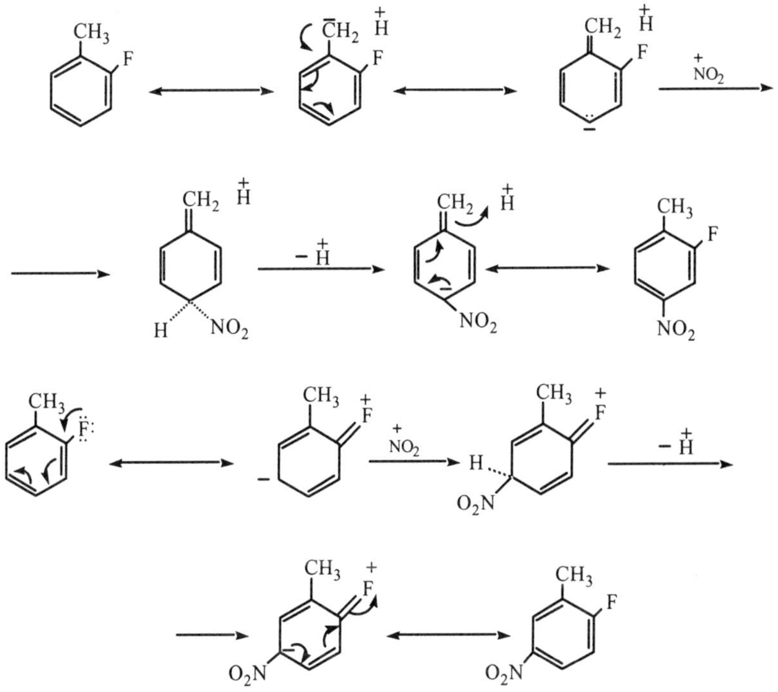

Alkyl groups activate the aromatic rings by increasing the electron density of the ring, whereas halogens deactivate the aromatic rings because of inductive decrease of electron density. However, fluorine is an exception. Because of its strong back-donation of electrons, it activates the benzene ring more strongly than the methyl group, as evidenced by the result of the nitration. The nitro group enters predominantly position *para* to fluorine, and the main product, **U**, is 2-fluoro-5-nitrotoluene [*43*].

7 Reactions of Sulfur Trioxide

⇨ EXPLANATION 26

Ethyl pentafluoroisopropenyl ether reacts in its resonance structure with sulfur trioxide in such a way that the sulfur atom of sulfur trioxide becomes attached to the terminal difluoromethylene group. The oxygen atom of sulfur trioxide combines with the methylene carbon of the ethyl group, and thus ethyl 2-oxopentafluoropropane-1-sulfonate is formed. Its transesterification with trifluoroacetic acid affords 2-oxopentafluoropropanesulfonic acid and ethyl trifluoroacetate [*44*].

⇨ EXPLANATION 27

Sulfur trioxide can add to the triple bond of perfluoro-*tert*-butylacetylene in two directions. Since it has only six electrons in the valence shell, it acts as a Lewis acid and combines with the more electronegative end of the triple bond. In perfluoro-*tert*-butylacetylene, the acetylenic fluorine increases the electron density at the vicinal carbon by back-donation of electrons, while the perfluoro-*tert*-butyl group increases the electronegativity at the same carbon by inductive electron withdrawal. It is sulfur of the sulfur trioxide that becomes attached to the more electronegative (electrophilic) carbon, the carbon adjacent to the perfluoro-*tert*-butyl group. The four-membered cyclic intermediate thus formed rearranges to perfluoro-*tert*-butylketene-α-sulfonyl fluoride [45].

$$(CF_3)_3CC \equiv CF \longleftrightarrow (CF_3)_3C\overset{-}{C} = \overset{+}{C}F \xrightarrow{O=S=O} (CF_3)_3CC = C-F \longrightarrow (CF_3)_3CC = C=O$$

EXPLANATION 28

As in Explanation 27, sulfur of sulfur trioxide becomes attached to the more electronegative (electrophilic) carbon of the double bond of perfluoroisobutylene. The electron density of the central carbon is increased by back-donation of electrons of the difluoromethylene group, and by electron withdrawal of the two trifluoromethyl groups. The intermediate four-membered cyclic sulfate rearranges to 2,2-bis(trifluoromethyl)-1-fluoroethenyl fluorosulfate [46].

$$(CF_3)_2C = CF_2 \longleftrightarrow (CF_3)_2\overset{-}{C} - \overset{+}{C}F_2 \longrightarrow (CF_3)_2C \overset{F}{-} C-F \longrightarrow (CF_3)_2C = CF$$

19%   X   15%   Y

When an excess of sulfur trioxide is used, two molecules of sulfur trioxide add to the double bond of perfluoroisobutylene in such a way that the sulfur atom becomes attached to the more electrophilic central carbon of the double bond to form a six-membered cyclic disulfate that rearranges to 2,2-bis(trifluoromethyl)-1-fluoroethenyl fluoropyrosulfate [46].

$$(CF_3)_2C = CF_2 + SO_3 \text{ (excess)} \xrightarrow{170-190°} (CF_3)_2C \overset{F}{-} C-F \longrightarrow (CF_3)_2C = CF$$

39%   Z   A   24%

EXPLANATION 29

Sulfur trioxide opens the epoxide ring by attaching its oxygen atom to the sterically more accessible difluoromethylene carbon. The propylene oxide oxygen combines with the sulfur atom and forms a five-membered cyclic sulfate [47].

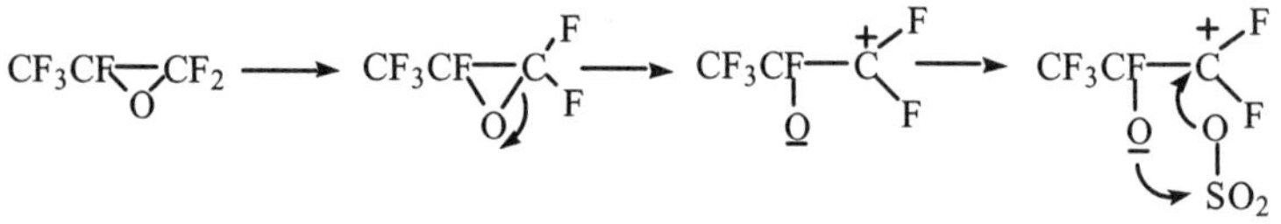

Simultanously, its isomer, 2-oxo-1,1,3,3,3-pentafluoropropyl fluoro-sulfate, is formed by a rearrangement.

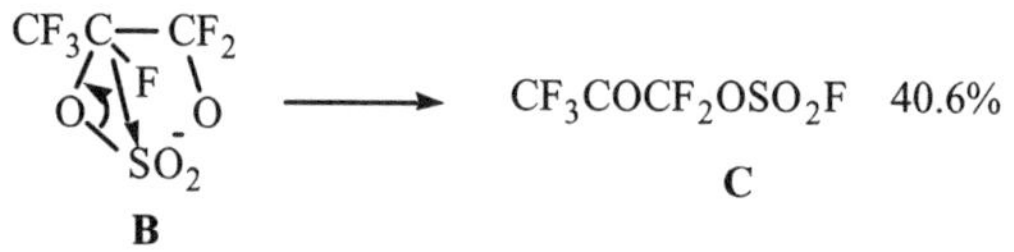

EXPLANATION 30

This peculiar reaction seems to be an insertion of sulfur trioxide between the carbon and fluorine atoms in the trifluoromethyl group. The product is perfluorobenzyl fluorosulfate [48, 49].

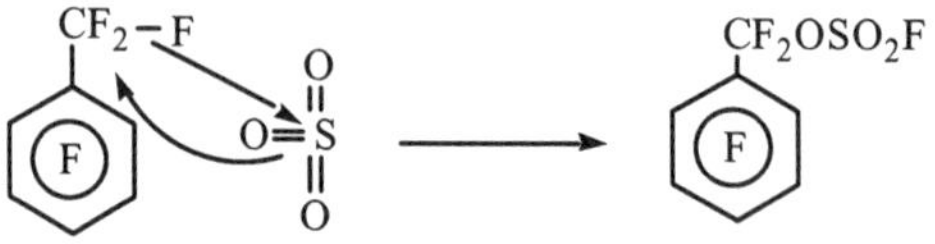

<table>
<tr><td>8</td><td>

Acid-Catalyzed Additions and Substitutions

</td></tr>
</table>

EXPLANATION 31

The first step in the Friedel-Crafts synthesis is formation of an electro-phile capable of electrophilic substitution in the aromatic ring. Such an electrophile (or cation) is generated from an alkyl halide by its reaction with a Lewis acid, in this case boron trifluoride. Boron trifluoride abstracts one of the halogens from the molecule of 1-chloro-2-fluoropropane to form complex anions $\overline{B}ClF_3$ or $\overline{B}F_4$. Because the bond between boron and fluorine is one of the strongest of all chemical bonds, boron trifluoride abstracts fluorine because the formation of tetrafluoroborate anion decreases the enthalpy of the reaction by 139 kJ (33 kcal/mol). Formation of $\overline{B}ClF_3$ would decrease the enthalpy only by 60 kJ (14 kcal/mol). Despite the fact that the bond dissociation energy of carbon-fluorine bond is higher than that of carbon-chlorine bond (443 vs. 328 kJ, or 106 vs. 78 kcal/mol), the enthalpy of tetrafluoroborate anion is lower than that of chlorotrifluoroborate anion: 582 vs. 388 kJ (139 vs. 93 kcal/mol). Consequently, the electrophile formed from the chlorofluoropropane is isopropyl (and not propyl) cation, and the product is 1-chloro-2-phenylpropane [50].

$$CH_2ClCHFCH_3 + BF_3 \xrightarrow{-10° \text{ to } 10°} CH_2Cl\overset{+}{C}HCH_3 \xrightarrow{C_6H_6} C_6H_5\underset{\underset{CH_3}{|}}{C}HCH_2Cl \quad \mathbf{D}$$
$$+ \overline{B}F_4$$

➠ EXPLANATION 32

The reaction of benzene with 2-chloro-1,1,1-trifluoropropane starts with the formation of a complex of the alkyl halide and aluminum chloride. Abstraction of chlorine by aluminum chloride to form tetrachloroaluminate anion leaves a carbocation with a positive charge on the carbon adjacent to the trifluoromethyl group. The positive charge next to a strongly electronegative trifluoromethyl cluster is not very happy. Therefore, proton from the methyl group migrates to the adjacent carbon and generates the carbocation with the positive charge on the more distant terminal carbon. As a result, the product $\mathbf{E}$ is 1-phenyl-3,3,3-trifluoropropane [51].

$$\underset{CF_3CHCH_3}{\overset{\overset{Cl}{|}}{}} + AlCl_3 \longrightarrow [CF_3\overset{+}{C}HCH_3] \,\overline{A}lCl_4$$

$$CF_3\overset{+}{C}HCH_2H \longrightarrow CF_3CH_2\overset{+}{C}H_2$$

$$CF_3CH_2\overset{+}{C}H_2 + C_6H_6 \longrightarrow CF_3CH_2CH_2C_6H_5 \quad \mathbf{E}$$

If the nonfluorinated compound 2-chloropropane reacted with benzene under similar conditions, the product would be isopropylbenzene.

$$CH_3CHClCH_3 + C_6H_6 \xrightarrow{AlCl_3} \begin{array}{c} CH_3CHCH_3 \\ | \\ C_6H_5 \end{array}$$

➥ EXPLANATION 33

In this Friedel-Crafts synthesis, benzene undergoes electrophilic substitution by a carbocation generated from an alkene and hydrogen fluoride. The trifluoromethyl group polarizes the double bond so that the negative charge is on the carbon adjacent to the trifluoromethyl group. This carbon is protonated, and the more distant carbon becomes attached to the benzene ring [52] .

$$CF_3CH{=}CH_2 \longleftrightarrow \overset{\vartheta-}{CF_3CH}{=}\overset{\vartheta+}{CH_2} \longleftrightarrow CF_3\overset{-}{CH}{-}\overset{+}{CH_2} \xrightarrow{\overset{+}{H}}$$

$$CF_3CH_2\overset{+}{CH_2} \xrightarrow{C_6H_6} CF_3CH_2CH_2C_6H_5 \quad \mathbf{F}$$

The parent compound, propylene, would react with benzene at the central carbon, because the protonation with the acid takes place at the carbon carrying two hydrogens, and the product would be isopropylbenzene.

➥ EXPLANATION 34

First, aluminum chloride abstracts the reactive benzylic fluorine from 1-phenylperfluoropropene and leaves a positively charged species that cyclizes to an indene derivative. The reactive benzylic fluorine atoms in this intermediate are replaced by chlorine [53] in a way similar to the replacement of fluorine by halogens in Explanations 22, 23, and 24.

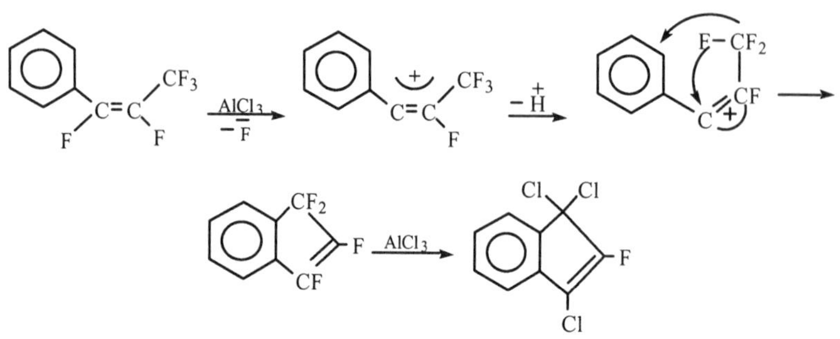

➡ EXPLANATION 35

In the hydrolysis of benzyl chloride, the leaving group is chloride anion. When benzyl fluoride is hydrolyzed in alkaline or neutral media, the leaving group is fluoride anion, which is a worse leaving group than chloride anion because the carbon-fluorine bond is much stronger than the carbon-chlorine bond: 443 kJ (106 kcal/mol) vs. 328 kJ (78 kcal/mol), respectively.

In acidic media, fluorine is protonated via hydrogen bonding, and thus the leaving group is not the fluoride ion but electrically neutral molecule of hydrogen fluoride, which is a much better leaving group than the negatively charged fluoride or chloride ions [54].

$$C_6H_5CH_2F \xrightarrow{HClO_4} C_6H_5CH_2\overset{+}{F}\text{---}H \xrightarrow{\overset{-}{OH}} C_6H_5CH_2OH + HF$$

Rate of Hydrolysis in 10% Aq. Acetone at 50° (k_0 x 10^6 sec^{-1})

Concentration of HClO$_4$ Benzyl Chloride   Benzyl Fluoride

➡ EXPLANATION 36

Hydrolysis of ω-fluoronitriles in alkaline media is an S_N2 reaction, which is very slow. In contrast, ω-fluorocarboxylic acids are hydrolyzed by an S_Ni mechanism because in alkaline media, five- and six-membered lactones are temporarily formed. The driving force is the strong tendency to close six- and especially five-membered rings [55].

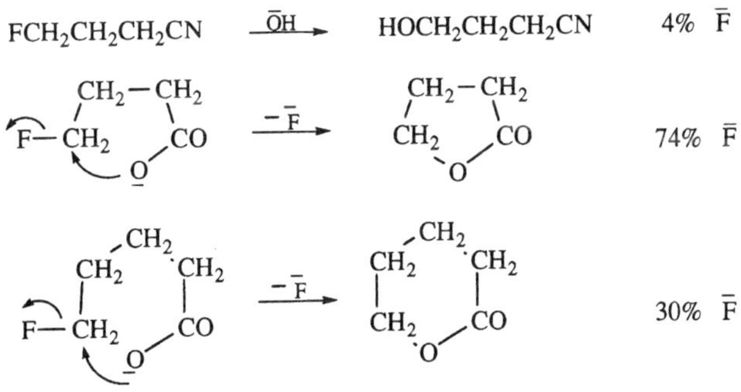

$$FCH_2CH_2CH_2CN \xrightarrow{\overset{-}{OH}} HOCH_2CH_2CH_2CN \qquad 4\% \;\overset{-}{F}$$

74% $\overset{-}{F}$

30% $\overset{-}{F}$

The first step in the hydrolysis in *tert*-butyl alcohol is probably nucleophilic displacement of fluorine in the reactive allylic position. Spontaneous elimination of hydrogen fluoride from the geminal fluorohydroxy compound yields perfluorocyclopent-2-en-1-one.

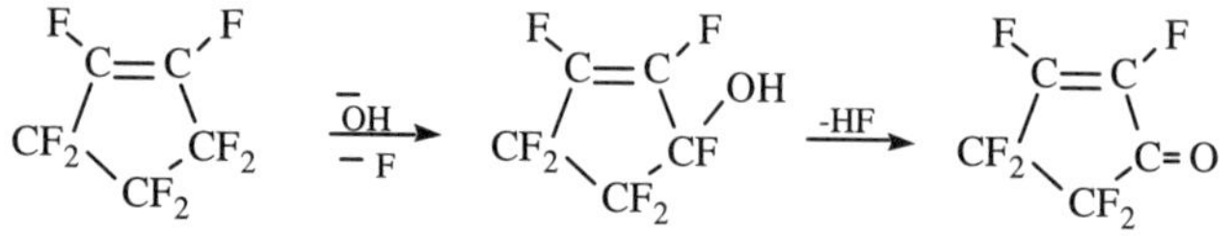

The next stage is an attack by hydroxide at the carbonyl carbon, followed by breaking of the bond between the carbonyl and the double bond. This reaction takes place in aqueous *tert*-butyl alcohol. (Primary and secondary alcohols as solvents would give different compounds, ethers.) The final product of the reaction is 5H-perfluoro-4-pentenoic acid (compound **G**) [56] .

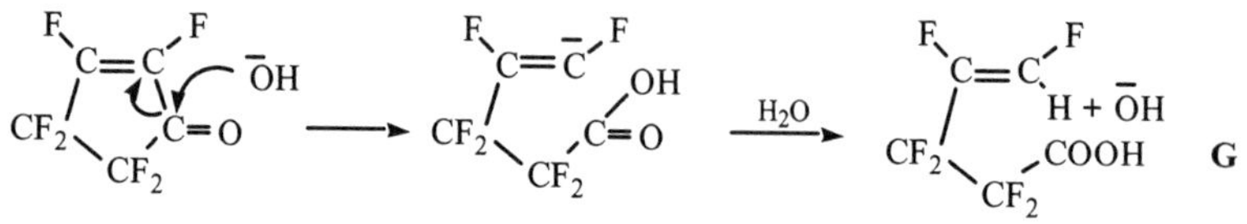

The reaction in diglyme starts with a nucleophilic displacement of fluorine at the double bond, whose electron density is decreased by the vicinity of strongly electronegative difluoromethylene groups.

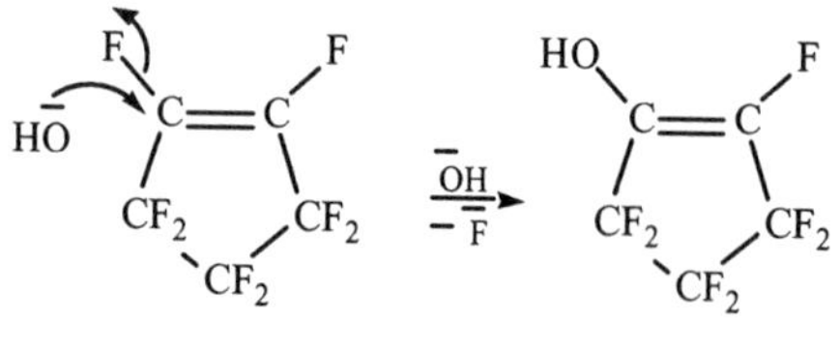

Nucleophilic displacement of fluorine atoms in the reactive allylic position affords an enol of a fluorinated 1,3-diketone, 2H-pentafluorocyclopentane-1,3-dione (compound **H**) [57] .

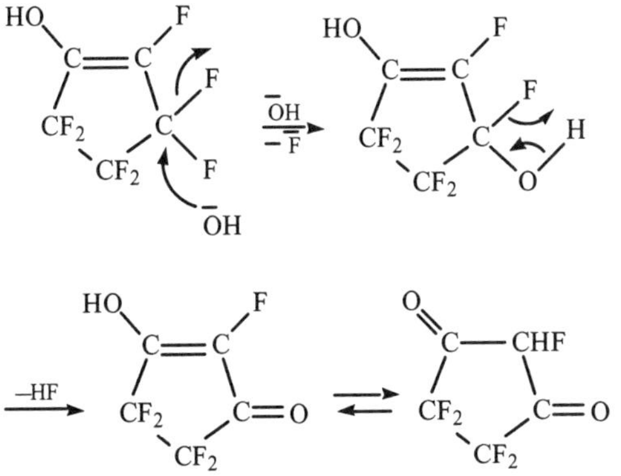

⟿ EXPLANATION 38

Because of the presence of nitrogen in the aromatic ring, electrons in pyridine are distributed in such a way that their density is higher in positions 3 and 5 (the β-positions). In these positions, electrophilic substitutions such as halogenation, nitration, and sulfonation take place. On the contrary, positions 2, 4, and 6 (α- and γ-positions, respectively) have lower electron density and are therefore centers for nucleophilic displacements such as hydrolysis or Chichibabin reaction. In the case of 3,5-dichlorotrifluoropyridine, hydroxide anion of potassium hydroxide attacks the α- and γ-positions because, in addition to the effect of the pyridine nitrogen, fluorine atoms in these position facilitate nucleophilic reaction by decreasing the electron density at the carbon atoms to which they are bonded. In a rate-determining step, hydroxyl becomes attached to the carbon atoms linked to fluorine and converts the aromatic compound into a nonaromatic "Meisenheimer" complex (see Surprise 67). To restore the aromaticity, fluoride ion is ejected in a fast step, and hydroxy pyridines **I** and **J** are obtained as the products [58].

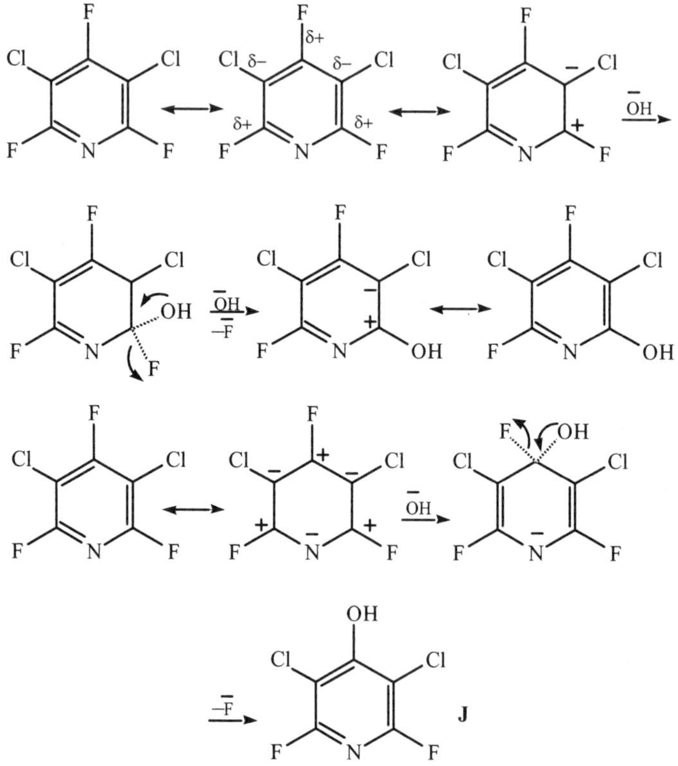

EXPLANATION 39

The presence of an amino group (or hydroxy group) in *ortho* (or *para*) position to a substituent increases the electron density in the aromatic ring:

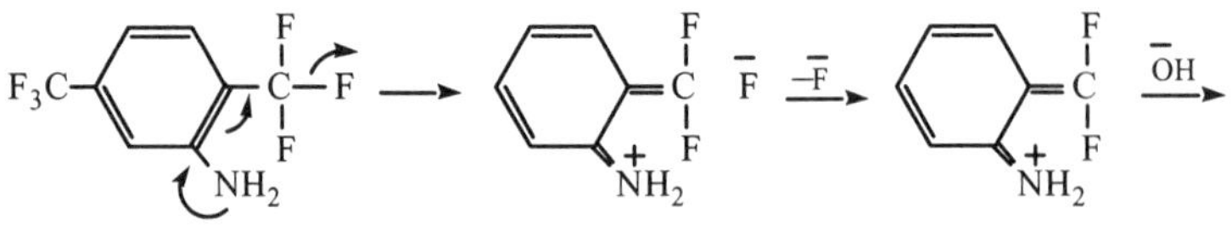

One fluorine atom is leaving as a fluoride ion, and the exocylic double bond to the difluoromethylene group is attacked by a hydroxide anion. Elimination of hydrogen fluoride gives an acyl fluoride that is readily hydrolyzed to the final product, 2-amino-4-trifluoromethylbenzoic acid (**K**) [59].

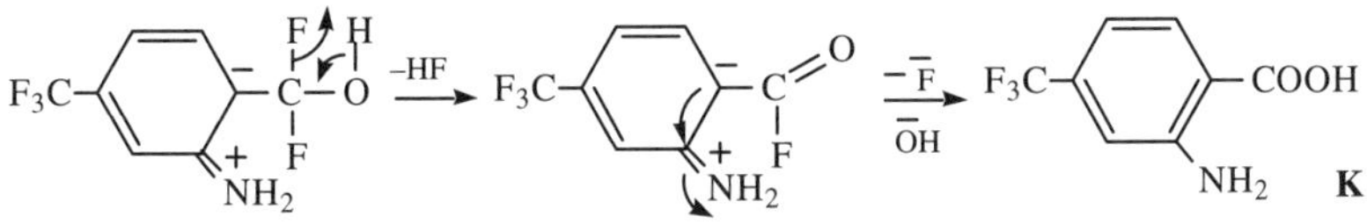

Because of the activation of the benzene ring by electron-donating substituents, hydrolysis of the trifluoromethyl group linked to the aromatic ring occurs easier than without such assistance.

EXPLANATION 40

The strong inductive effect of the heptafluoropropyl group facilitates the reaction of the compound with hydroxide anion at the aldehyde carbon. In an S_N2 reaction, the leaving group is heptafluoropropyl with its electron pair. Protonation of the anion gives heptafluoropropane **L** and potassium formate [60].

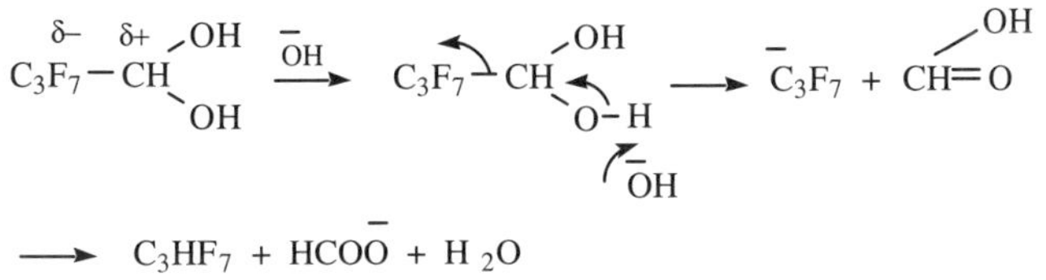

The same product is obtained from heptafluoropropyl methyl ketone [61]. The reaction resembles the fluoroform reaction of fluorinated alcohols [62] and ketones [63].

$$C_3F_7COCH_3 \xrightarrow[\text{50-60}°]{\text{1.2 } N \text{ NaOH}} C_3HF_7 + CH_3CO_2H \qquad [61]$$

$$C_6H_5COCF_3 \xrightarrow[]{\text{10\% KOH}} C_6H_5CO_2H + CHF_3 \qquad [62]$$

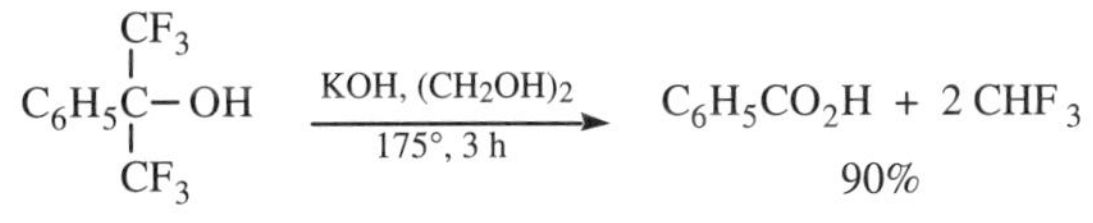

$$\underset{\underset{CF_3}{|}}{\overset{\overset{CF_3}{|}}{C_6H_5C}}\!-OH \xrightarrow[\text{175°, 3 h}]{\text{KOH, (CH}_2\text{OH)}_2} C_6H_5CO_2H + 2\,CHF_3 \qquad [63]$$
$$\qquad\qquad\qquad\qquad\qquad\qquad\qquad 90\%$$

EXPLANATION 41

Perfluoroalkyl iodides do not act as alkylating agents as the nonfluorinated iodides do. The strong inductive effect of the perfluoroalkyl group polarizes the carbon-iodine bond in an opposite direction to that in the nonfluorinated iodides. The carbon-iodine bond is cleaved to perfluoro-alkyl anion and iodine cation. Reaction with water yields pentafluoroet-hane and hypoiodic acid, with potassium hydroxide pentafluoroethane and potassium hypoiodide [64].

$$CH_3CH_2-I \; \longleftrightarrow \; \overset{\delta+}{CH_3CH_2}-\overset{\delta-}{I} \; \longrightarrow \; \overset{+}{CH_3CH_2} + \overset{-}{I}$$

$$CF_3CF_2-I \; \longleftrightarrow \; \overset{\delta-}{CF_3CF_2}-\overset{\delta+}{I} \; \longrightarrow \; \overset{-}{CF_3CF_2} + \overset{+}{I}$$

$$\overset{-}{CF_3CF_2} + H_2O \longrightarrow CF_3CHF_2 + \overset{-}{OH}; \quad \overset{+}{I} + \overset{-}{OH} \longrightarrow IOH$$

M

➥ EXPLANATION 42

The first reaction step is most probably hydrolysis of the aryl methyl ether by hydrobromic and acetic acid. Such a mixture is very efficient because one of the products, methanol, is continuously removed from the reaction mixture by esterification with acetic acid, and thus the reaction equilibrium between the starting material and the product is shifted toward the product.

The phenolic hydroxyl, resulting from the hydrolysis of the phenol ether, affects the distribution of electrons in the benzene ring. Back-dona-tion of electrons from the phenolic oxygen facilitates elimination of ben-zylic fluorine next to the benzene ring.

Because of the electron withdrawal of the trifluoromethyl group, the α,β-double bond is polarized so that the α-carbon with respect to the tri-fluoromethyl group has a negative charge, and the β-carbon a positive charge. The β-carbon is thus fit to accept a nucleophile, water, whose hydroxyl is a better nucleophile than are the bromide or acetate anion. Redistribution of the double bonds gives the enol form of the product, 1-(o-hydroxyphenyl)-2,3,3,3-tetrafluoro-1-propanone [65].

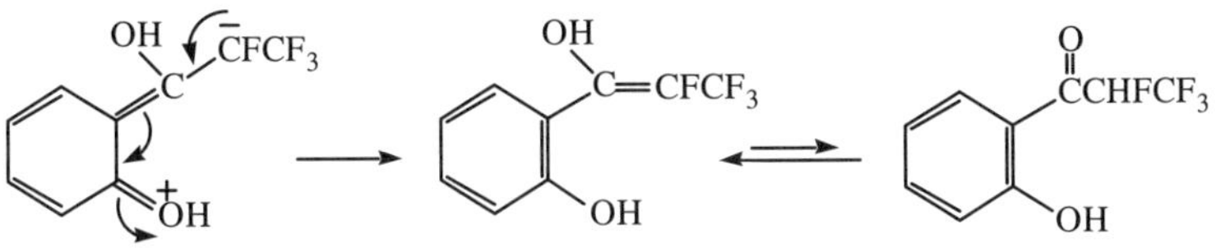
OH
CFCF3
OH
OH
C=CFCF3
OH
O
CCHFCF3
OH

EXPLANATION 43

The reaction in Explanation 43 is not a typical hydrolysis of geminal difluorides. A simple example of hydrolysis of geminal difluorides is conversion of 1,1-difluorocyclohexane to cyclohexanone. It is not easy, and takes place in alkaline or acidic media under very energetic conditions [*66*].

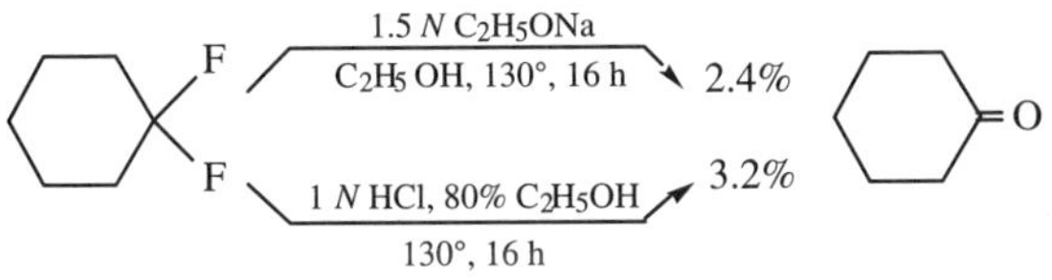
F
F
1.5 N C2H5ONa
C2H5 OH, 130°, 16 h
2.4%
1 N HCl, 80% C2H5OH
130°, 16 h
3.2%
O

The hydrolysis of the compound shown here is a stepwise reaction starting with dehydrofluorination by piperidine. The resulting α,β-unsaturated compound undergoes a nucleophilic addition of water, and the subsequent elimination of hydrogen fluoride affords a keto ester. Because the keto group is flanked by a perfluorohexyl cluster that stabilizes hydrates, the original difluoromethylene group is converted to a geminal dihydroxy derivative [*67*].

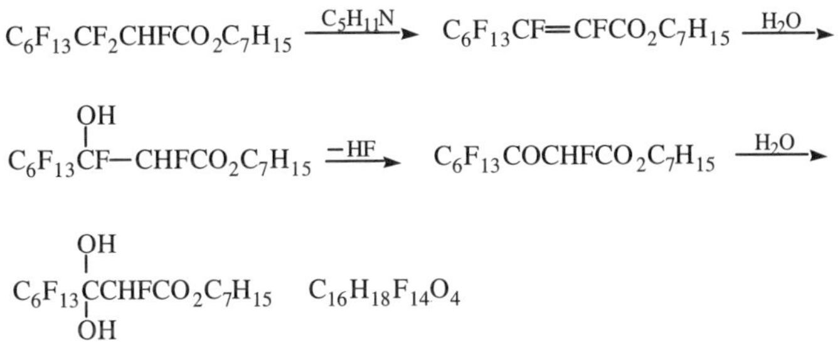
C6F13CF2CHFCO2C7H15
C5H11N
C6F13CF=CFCO2C7H15
H2O
OH
C6F13CF—CHFCO2C7H15
−HF
C6F13COCHFCO2C7H15
H2O
OH
C6F13CCHFCO2C7H15
OH
C16H18F14O4

EXPLANATION 44

Surprisingly, heating of the enol ether 3,3-difluoro-2-ethoxy-1-phenyl-1-cyclobutene with concentrated sulfuric acid does not affect the ether bond, but results in the hydrolysis of the difluoromethylene group to a keto group. The reason for this reaction is probably the high reactivity of the allylic fluorines. Protonation of the first fluorine via a hydrogen bond facilitates its ejection, and a nucleophilic attack of the carbocation by water gives an unstable geminal fluorohydroxy compound that eliminates hydrogen fluoride and proton [68].

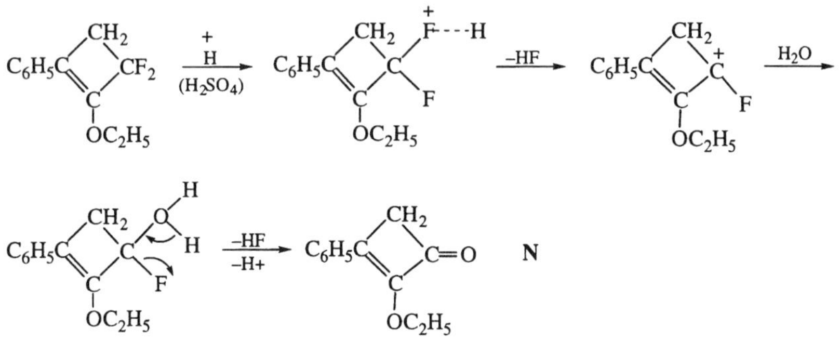

EXPLANATION 45

The reaction starts probably with dehydrofluorination to give an α,β-unsaturated ketone. This undergoes a nucleophilic addition of water, and a loss of hydrogen fluoride from the intermediate affords a β-diketone.

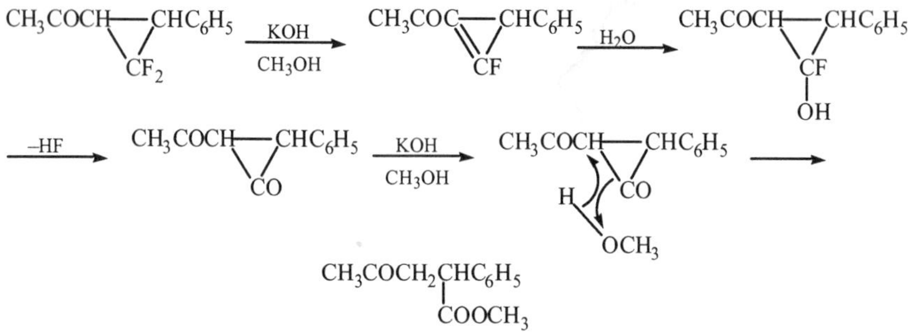

The bond between the two carbonyl groups of the β-diketone is cleaved by potassium methoxide present in the methanolic aqueous potassium hydroxide, and the cyclopropane ring is split open while the carbonyl group is converted to a carboxyl to give methyl 4-oxo-2-phenylpentanoate [69].

EXPLANATION 46

In alkaline media, phenol loses its proton, and is converted to a phenoxide ion (I):

Back-donation of electrons by oxygen changes formula I to its resonance hybride (II), which by a loss of fluoride ion affords a quinonoid
compound (III). Addition of water across the exocyclic double bond
gives compound IV, which ejects a molecule of hydrogen fluoride and
gives acyl fluoride (V). Hydrolysis of the acyl fluoride yields the corresponding carboxylic acid (VII), and allylic shift of hydrogen changes VII
to *p*-hydroxybenzoic acid (VIII) [70].

10 Alkylations

⟵ EXPLANATION 47

The clue to the reaction is the polarity of the double bond. The lowest
electron density is at the carbon linked to hydrogen, and is due to the
strong inductive effect of the difluoromethylene groups and a slight
effect of the vinylic chlorine. The attacking species, ethoxide anion,
which is in an equilibrium with hydroxide ion in alcoholic potassium
hydroxide, reacts in an S_N2 reaction by joining the carbon bonded to
hydrogen. The subsequent shift of the double bond facilitates ejection of
fluorine as an anion and leads to an ether, compound **O** [71].

In the presence of an excess of potassium ethoxide, the ethoxide anion
joins the less negative end of the double bond. Strong electron back-
donating power of fluorine increases negativity at the carbon linked to
chlorine and causes the addition of the ethoxide ion to the carbon linked
to fluorine. Subsequent ejection of fluoride anion restores the double
bond, and the final product is a diether, compound **P** [71].

$$CF_2\text{-}CH \xrightarrow{} {}^-OC_2H_5$$
$$\underset{F\text{-}CF\text{-}CCl}{\overset{\delta+}{}} \quad \xrightarrow{-\bar{F}} \quad \underset{CF=CCl}{CF_2\text{-}CHOC_2H_5} \qquad \mathbf{O}$$

$$\underset{F\text{-}C=CCl}{CF_2\text{-}CHOC_2H_5} \xrightarrow{C_2H_5\bar{O},\ H_2O} \underset{\underset{OC_2H_5}{F\text{-}C\text{-}CCl}}{CF_2\text{-}CHOC_2H_5} \xrightarrow{-\bar{F}} \underset{\underset{OC_2H_5}{C=CCl}}{CF_2\text{-}CHOC_2H_5} \qquad \mathbf{P}$$

✎ EXPLANATION 48

In ethanolic solution of potassium hydroxide, ethoxide ions are in an equilibrium with hydroxide ions. Because the ethoxide ion is more nucleophilic than the hydroxide ion, it is the reactive species in the reaction with the bromochlorotetrafluorocyclobutene. Since the resonance effect of chlorine is slightly stronger than that of bromine, the double bond is polarized in such a way that the carbon bonded to bromine has slightly higher electron density than the carbon bonded to chlorine. Consequently, the ethoxide anion joins predominantly the carbon linked to chlorine. The intermediate carbanion is protonated at the carbon carrying bromine, and gives 2-bromo-1-chloro-1-ethoxytetrafluorocyclobutane, which ejects hydrogen chloride and affords 2-bromo-1-ethoxytetrafluorocyclobutene **Q**. The same product could be formed by ejection of the chloride anion from the primarily generated carbanion.

$$\underset{\underset{\delta+\delta-}{Cl\text{-}C=CBr}}{CF_2\text{-}CF_2} \xrightarrow{C_2H_5\bar{O}} \underset{\underset{\underset{-Cl}{OC_2H_5}}{Cl\text{-}C\text{-}CBr}}{CF_2\text{-}CF_2} \xrightarrow{\overset{+}{H}} \underset{\underset{OC_2H_5}{Cl\text{-}C\text{-}CHBr}}{CF_2\text{-}CF_2} \xrightarrow{-HCl} \underset{\underset{OC_2H_5}{C=CBr}}{CF_2\text{-}CF_2} \quad \mathbf{Q}$$

However, because of a small difference in the polarity of the double bond, the ethoxide anion joins also the carbon linked to bromine, and the result is 2-chloro-1-ethoxytetrafluorocyclobutene, compound **R**, formed in a smaller amount than compound **Q** [72].

$$\underset{CCl=C\text{-}Br}{CF_2\text{-}CF_2} \xrightarrow{C_2H_5OH} \underset{\underset{OC_2H_5}{ClCH\text{-}C\text{-}Br}}{CF_2\text{-}CF_2} \xrightarrow{-HBr} \underset{\underset{OC_2H_5}{CCl=C}}{CF_2\text{-}CF_2} \quad \mathbf{R}$$

The first reaction stage is undoubtedly S_N2 nucleophilic displacement of chlorine by the ethyl acetoacetate anion. Chlorine bond to carbon is weaker than that of fluorine. The resulting ethyl 5-fluoropentan-2-one-3-carboxylate subsequently cyclizes by an S_Ni reaction, displacing this time the fluorine atom [73].

$$CH_3COCH_2CO_2C_2H_5 \xrightarrow{C_2H_5\bar{O}} CH_3CO\bar{C}HCO_2C_2H_5 + C_2H_5OH$$

$$FCH_2CH_2Cl + \bar{C}H\underset{CO_2C_2H_5}{\overset{COCH_3}{\diagdown}} \xrightarrow{-\bar{Cl}} FCH_2CH_2CH\underset{CO_2C_2H_5}{\overset{COCH_3}{\diagdown}} \xrightarrow{C_2H_5\bar{O}}$$

$$\underset{CH_2}{\overset{\overset{F}{|}}{\underset{|}{CH_2}}} \underset{\overset{|}{C}}{\overset{COCH_3}{\diagdown}} CO_2C_2H_5 \xrightarrow{-\bar{F}} \underset{CH_2}{\overset{CH_2}{|}}\underset{C}{\diagdown}\underset{CO_2C_2H_5}{\overset{COCH_3}{\diagup}} \quad (C_8H_{12}O_3)$$

Although cyclization to three-membered rings is not as easy and as frequent as are cyclizations to six- and especially five-membered rings, it is not so uncommon with fluorinated compounds. In this particular case, the voluminous carbanion may cyclize easier than would with chlorine substituent for steric reasons. Precedents of such cyclizations are formations of ethylene oxide compounds from fluorohydrins [74].

$$FCH_2\underset{OH}{\overset{|}{CHCH_2F}} \xrightarrow[\text{reflux 3 h}]{\text{KOH (solid)}} FCH_2CH\overset{}{\underset{O}{-}}CH_2 \quad 75\%$$

Chlorodifluoromethane and sodium hydroxide, by elimination of hydrogen chloride, form difluorocarbene, which is inserted between oxygen and hydrogen of pentafluorophenol. Such insertion is facilitated by high polarity of the oxygen-hydrogen bond owing to the strong electronegative effect of the pentafluorophenyl group.

Another possible mechanism is formation of pentafluorophenoxide anion that combines with difluorocarbene, and the intermediate anion is rapidly protonated.

$$\overset{\delta-\ \ \delta+}{C_6F_5O-H} + :C\overset{F}{\underset{F}{\diagdown}} \longrightarrow C_6F_5O-CHF_2$$

$$C_6F_5OH \xrightarrow[-H_2O]{\bar{O}H} C_6F_5\bar{O} \xrightarrow{:CF_2} C_6F_5O\bar{C}F_2 \xrightarrow{H_2O} C_6F_5CHF_2 + \bar{O}H$$

Hydrogen in α-position to the phenolic oxygen is fairly acidic. It is therefore not surprising that in an alkaline medium, hydrogen fluoride is eliminated and a new carbene is generated. It is subsequently inserted between the oxygen and hydrogen of a second molecule of pentafluorophenol, and thus compound **S**, bis(pentafluorophenoxy)fluoromethane, is formed [75].

$$C_6F_5OCHF_2 \xrightarrow[-HF]{NaOH} C_6F_5O\overset{..}{C}F \xrightarrow{C_6F_5OH} C_6F_5OCHFOC_6F_5 \qquad \mathbf{S}$$

⟜ EXPLANATION 51

The double bond in 1,4-dibromohexafluoro-2-butene surrounded by fluorines and difluoromethylene groups is prone to undergo a nucleophilic addition of ethanol. From the addition product, hydrogen bromide is eliminated in the alkaline medium by E2 mechanism, and the compound $C_6H_5BrF_6O$ is formed. Another explanation of the reaction is addition of ethoxide anion followed by elimination of bromide anion [76] .

$$BrCF_2CF{=}CFCF_2Br + C_2H_5OH \xrightarrow{C_2H_5ONa} \underset{\underset{OC_2H_5}{|}}{BrCF_2CFCHFCF_2Br}$$

$$\underset{\underset{OC_2H_5}{|}}{BrCF_2CF\overset{H\ \frown\ \bar{O}C_2H_5}{CF_2{-}Br}} \longrightarrow \underset{\underset{OC_2H_5}{|}}{BrCF_2CF{-}CF{=}CF_2} + \bar{B}r + C_2H_5OH$$

$$BrCF_2CF{=}CFCF_2Br \xrightarrow{C_2H_5\bar{O}} \underset{\underset{OC_2H_5}{|}}{BrCF_2CF{-}F{-}CF_2{-}Br} \longrightarrow \underset{\underset{OC_2H_5}{|}}{BrCF_2CFCF{=}CF_2} + \bar{B}r$$

⟜ EXPLANATION 52

Allylic fluorines in the starting compounds are reactive enough to be displaced, in an S_N2 reaction, by the methoxide anion. The allylic fluorine at the other end of the double bond is displaced by the oxygen of the methyl ether in an $S_N i$ reaction, and after ejection of methyl fluoride, a five-membered cyclic ether, perfluoro(tetramethyl)-2,5-dihydrofuran, is formed. Compounds **T** and **U** are shown below [77] .

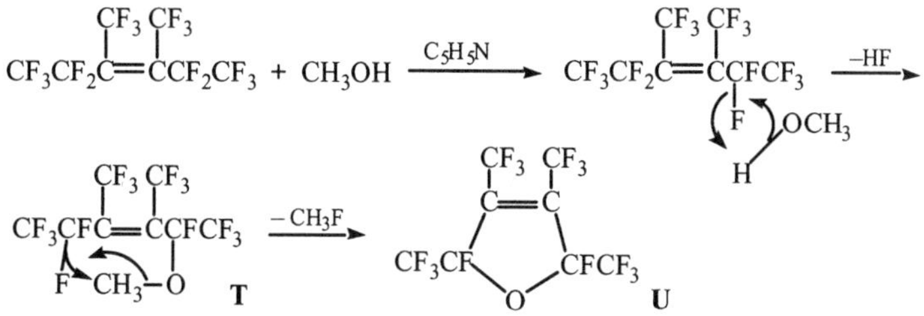

EXPLANATION 53

The double bond in chlorotrifluoroethylene is polarized by back-donation of electrons of fluorine in such a way that the negative charge is on the carbon linked to chlorine and fluorine. Consequently, the difluoromethylene end of the double bond is more electrophilic and is attacked by the ethoxide anion. Subsequent ejection of fluoride anion gives an unsaturated intermediate, 1-chloro-1,2-difluoro-2-ethoxyethylene, compound **V**. This compound reacts with another ethoxide anion in a similar way and yields 1-chloro-2,2-diethoxy-1-fluoroethylene. Nucleophilic addition of a third molecule of ethanol gives the final product, the orthoester of chlorofluoroacetic acid, compound **W** [78].

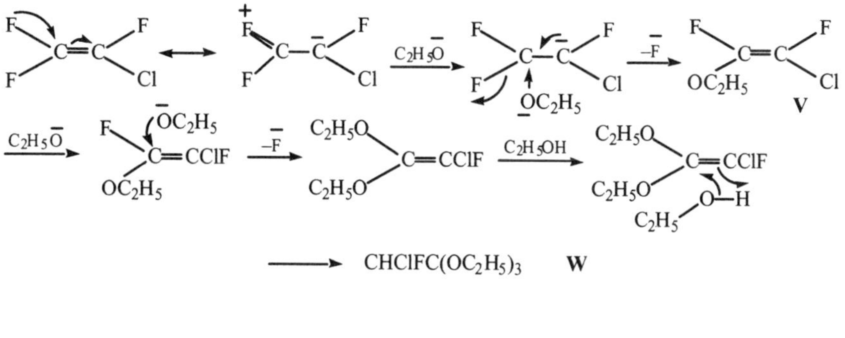

$$\longrightarrow \quad CHClFC(OC_2H_5)_3 \quad \textbf{W}$$

EXPLANATION 54

The reaction starts undoubtedly by opening of the lactone ring by cesium fluoride to form cesium salt of 4-hydroxyperfluorobutyryl fluoride. The anion of this intermediate opens the ring of perfluoropropylene oxide at the central carbon. Expulsion of fluoride anion from the terminal difluoromethylene group converts the intermediate to the difluoride of perfluoro-

2-methyl-3-oxaheptanedioic acid, compound **X** [79]. What is unexpected in this reaction is the nucleophilic attack at the central carbon rather than at the sterically more accessible difluoromethylene group [79].

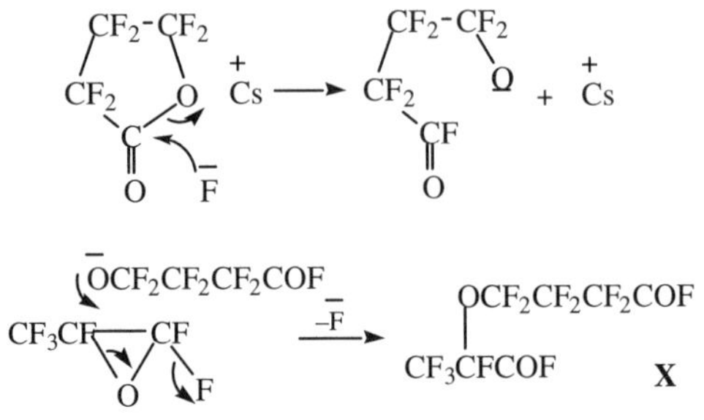

⟢ EXPLANATION 55

Unlike alkyl halides in which nucleophiles react with alkyl groups because the carbon-halogen bond is cleaved so as to form carbocation and a halide anion, in polyfluroalkyl halides the carbon-halogen bond is polarized in the opposite sense to form fluoroalkyl carbanion and positively charged halogen (see Surprises 40 and 41). Thus, the nucleophile reacts with the halogen rather than with the fluoroalkyl group.

In the case of thiophenoxide and dichlorodifluoromethane, such a reaction forms chlorodifluoromethyl carbanion, which dissociates to difluorocarbene and chloride anion. Difluorocarbene reacts with the thiophenoxide anion to form phenylthiodifluoromethyl anion. This unstable species reacts in two ways. (1) It abstracts chlorine from another molecule of dichlorodifluoromethane and forms chlorodifluoromethyl phenyl sulfide **Y**. (2) It reacts with hydrogen of water to form difluoromethyl phenyl sulfide **Y'**. Product **Y** can also react with two molecules of thiophenoxide anion to give bis(thiophenyl)difluoromethane **Y''** [80].

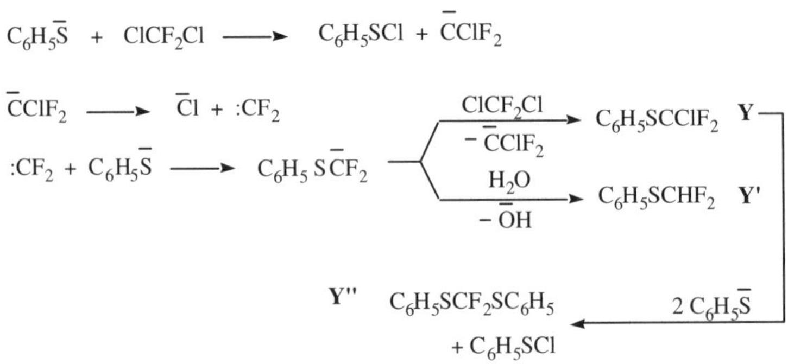

The reaction may also be explained by a free radical chain mechanism $S_{AR}1$, whose steps were confirmed by electron spin resonance (ESR).

$$C_6H_5\bar{S} + CF_3Br \longrightarrow C_6H_5\overset{\bullet}{S} + CF_3\bar{Br}\bullet$$

$$CF_3\bar{Br}\bullet \longrightarrow \overset{\bullet}{C}F_3 + \bar{Br}$$

$$C_6H_5\bar{S} + \overset{\bullet}{C}F_3 \longrightarrow C_6H_5\bar{S}CF_3\bullet$$

$$C_6H_5\bar{S}CF_3\bullet + CF_3Br \longrightarrow C_6H_5SCF_3 + CF_3\bar{Br}\bullet$$

⟹ EXPLANATION 56

The reaction starts probably by elimination of the tertiary fluorine atoms. Thus, a bicyclic olefin is formed. Its double bond activates the four allylic fluorines for nucleophilic displacement by thiophenoxide anions to give 1,4,5,8-tetrakis(phenylthio)perfluoro-$\Delta^{9,10}$-octalin.

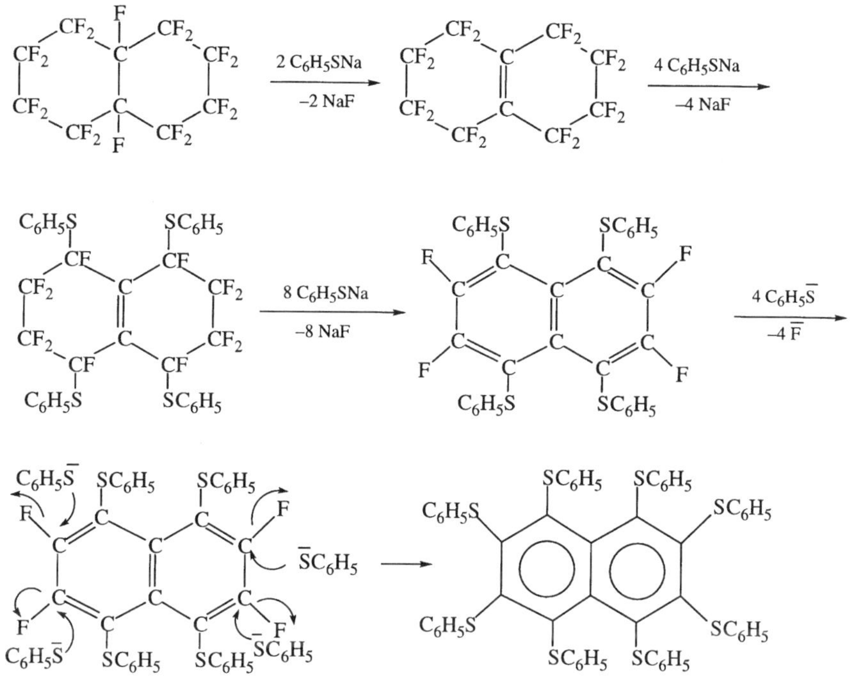

The next step could be defluorination of the tetrakis(phenylthio)octalin. Double bonds thus formed change the system to a derivative of naphthalene, 2,3,6,7-tetrafluoro-1,4,5,8-tetrakis(phenylthio)naphthalene. If this happens, it is a very unusual reaction, because sodium thiophenolate is not known to cause elimination of fluorine, and especially not under gentle conditions. Once the intermediate contains aromatic fluorines, their displacement by a strong nucleophile such as sodium thiophenoxide is quite understandable [81].

➡ EXPLANATION 57

Because of the strong inductive and resonance effects of the fluorine-rich environment, the double bond is strongly activated for nucleophilic additions:

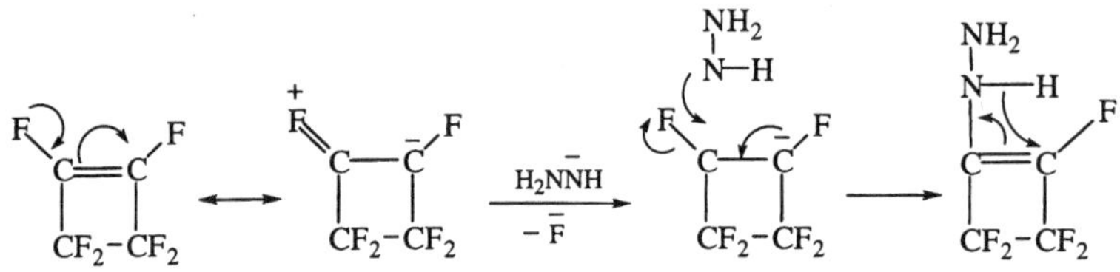

Fluorine atoms in allylic positions are very prone for nucleophilic displacement by the anions of hydrazine:

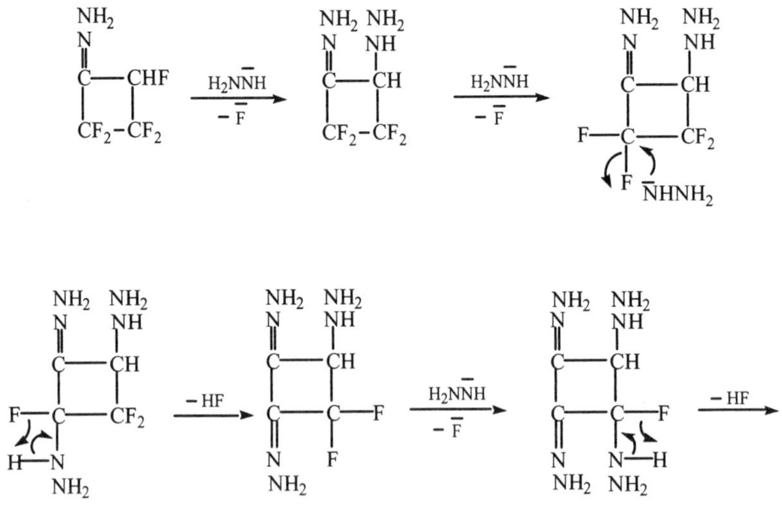

So far so good. But how can the fourth double bond be created? There is a possibility of dehydrogenation of –CH–NH– group by hydrazine, similar to that of dehydrogenation of –CH–OH bond in saccharides by phenylhydrazine, which is hydrogenolyzed to ammonia and aniline. Here the molecule of cyclobutanetetronetetrakis(hydrazone), **Z**, is obtained, and hydrazine is split to ammonia [82].

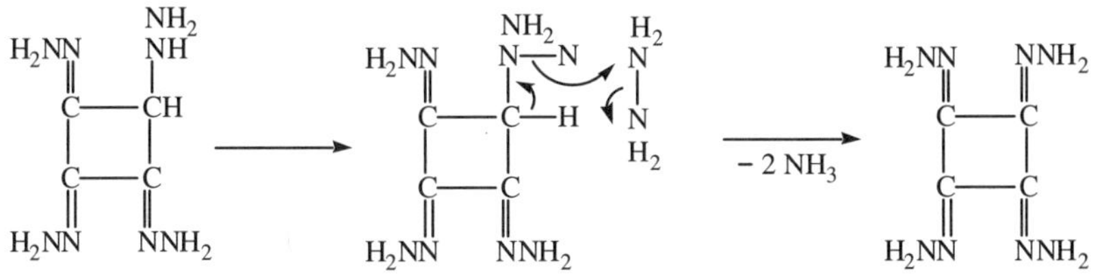

In the same way, cyclopentanepentonepentakis(phenylhydrazone) was prepared in 88% yield from 1,2-dichlorohexafluorocyclopentene [82].

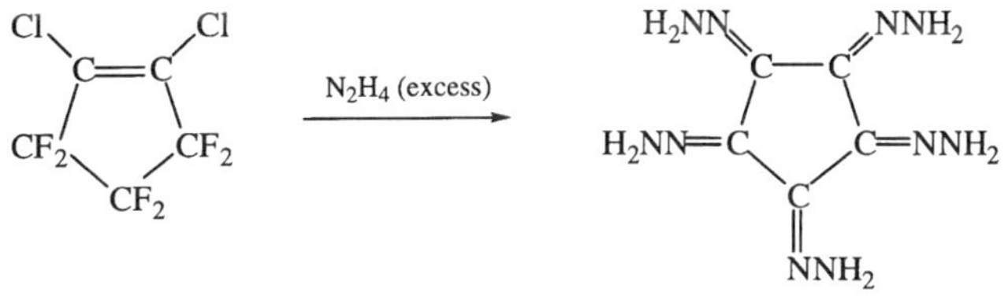

⟸ EXPLANATION 58

The clue to the reaction is the reactivity of benzylic fluorines. The trifluoromethyl group is gradually converted to an amidine group, and ultimately to a nitrile group [83]. Similar transformation affects two other trifluoromethyl groups. The trifluoromethyl group flanked by two trifluoromethyl groups escapes such a conversion, perhaps for steric reasons. The product of the reaction is **A**, 2,4,6-tricyanobenzotrifluoride.

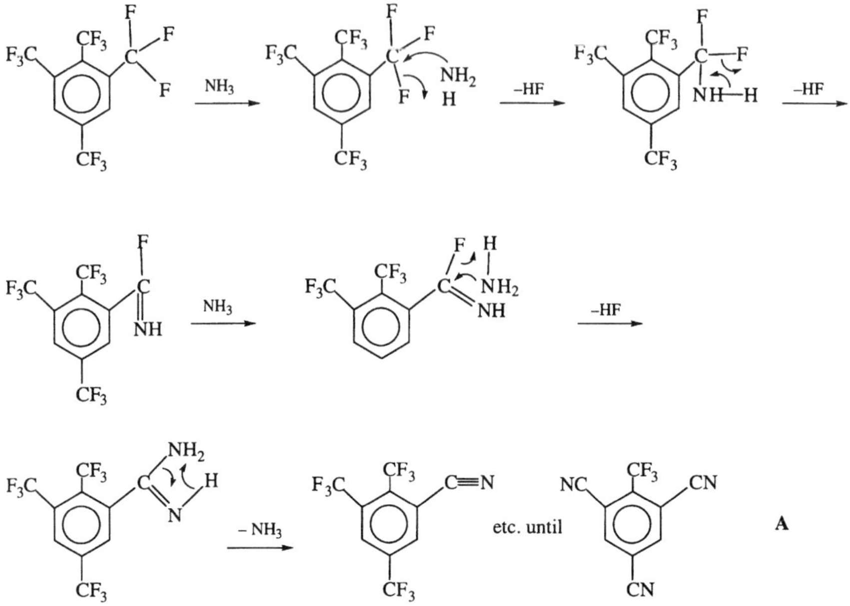

F₃C, CF₃, C, F, F, NH₃ → ... NH₂, H, −HF → ... NH—H, −HF →

F₃C, CF₃, C, NH, NH₃ → F₃C, CF₃, C, NH₂, NH, −HF →

F₃C, CF₃, C, NH₂, H, N, −NH₃ → F₃C, CF₃, C≡N, CF₃, etc. until, NC, CF₃, CN, CN, **A**

⟾ EXPLANATION 59

Because of the inductive and resonance effects of fluorine, the double bond has a very low electron density, especially at the terminal difluoro-methylene group. Dimethyl malonate adds nucleophilically across the double bond.

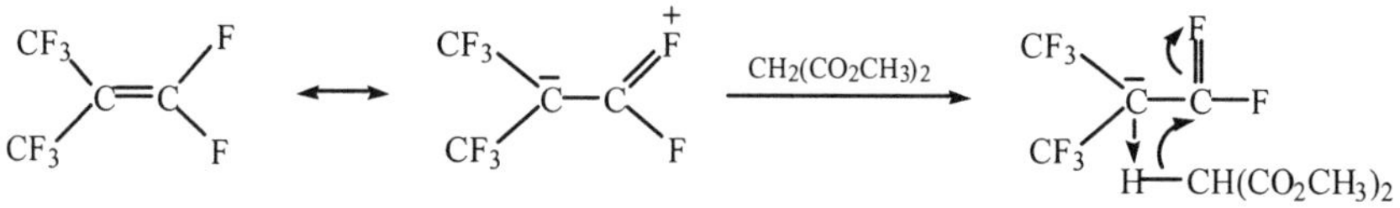

$CH_2(CO_2CH_3)_2$

CF_3 ... $C=C$ ... F ↔ CF_3 ... $C-C$ ... F (+) → CF_3 ... $C-C-F$ ... $H-CH(CO_2CH_3)_2$

Elimination of two molecules of hydrogen fluoride affords **B**, an allene derivative, methyl 2-methoxycarbonyl-5,5,5-trifluoro-4-trifluoromethyl-penta-2,3-dienoate [84].

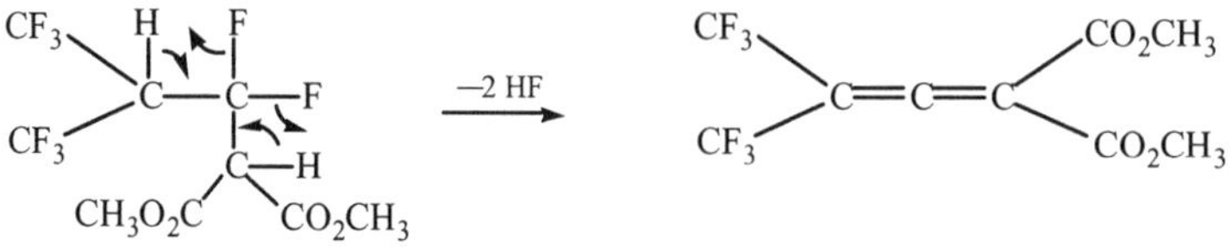

CF_3, H, F, C—C—F, CF_3, C—H, CH_3O_2C, CO_2CH_3 —2 HF→ CF_3, CF_3, $C=C=C$, CO_2CH_3, CO_2CH_3, **B** [84]

At 620°C, chlorodifluoromethane eliminates hydrogen chloride and generates difluorocarbene which is inserted between chlorine and carbon in the trichloromethyl group. Subsequent elimination of a molecule of chlorine gives the final product **C**, α-chloroperfluorostyrene [85].

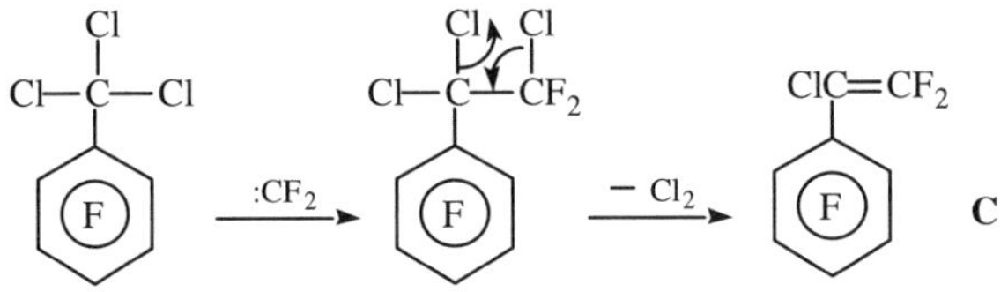

First, 2-phenylaminocarbonylcyclopentanone is treated with morpholine to give the corresponding enamine, 1-(*N*-morpholinyl)-5-phenylaminocarbonylcyclopentene.

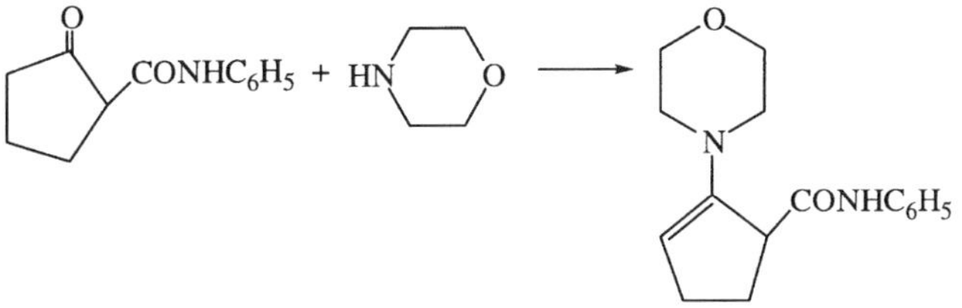

The enamine then reacts in the allylic position at carbon 5. This position is activated not only by the double bond, but also by the neighboring carbonyl group. Hydrogen atom in this position is fairly acidic and is easily replaced by a nucleophile, the trifluoromethanesulfenyl group. Elimination of hydrogen chloride affords 1-(*N*-morpholinyl)-5-phenylaminocarbonyl-5-trifluoromethanesulfenylcyclopentene. But there is still another reactive center in the molecule, the double bond. Trifluoromethanesulfenyl chloride adds to the double bond. Back-donation of electrons by nitrogen makes carbon 2 more electronegative, and that is where the sulfur becomes attached. Subsequent elimination of hydrogen chloride restores the double bond, and the final product is **D**, 1-(*N*-morpholinyl-5-phenylaminocarbonyl-2,5-bis(trifluoromethanesulfenyl)cyclopentene [86].

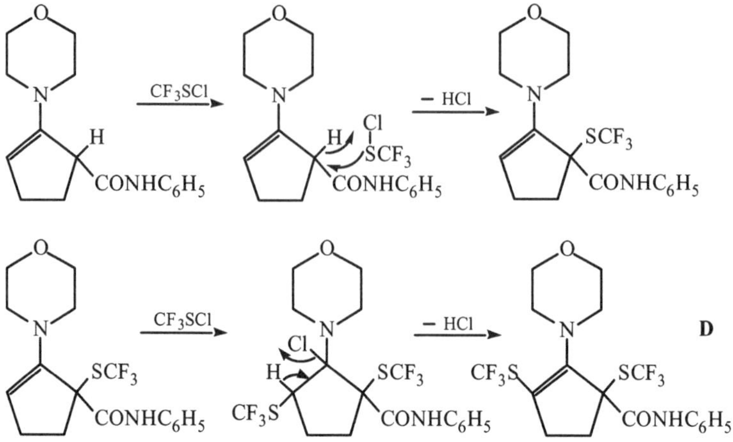

EXPLANATION 62

Sodium methoxide is a catalyst for addition of alcohols across double bonds of fluorinated olefins. The reagent in this reaction is methoxide anion, which adds to the carbon of the double bond that has lower electron density. Because of a strong electron π-back-donation by fluorine, the lower electron density is at the carbon linked to two fluorines.

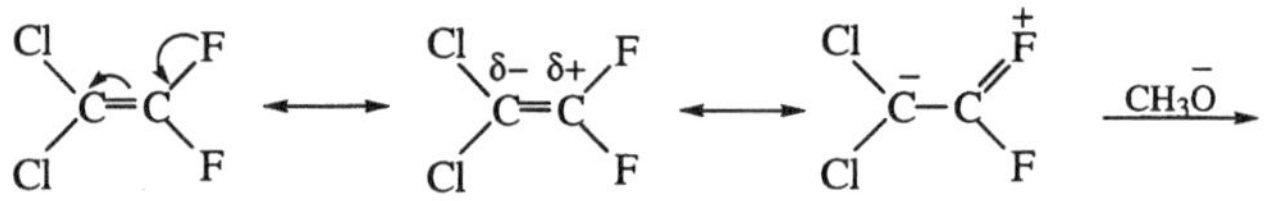

The methoxide anion joins the carbon of the difluoromethylene group and generates a carbanion that is protonated by methanol to afford **E**, 2,2-dichloro-1,1-difluoroethyl methyl ether and methoxide anion that perpetuates the reaction.

The intermediate carbanion may eject fluorine and thus afford an unsaturated ether (**F**), 2,2-dichloro-1-fluorovinyl methyl ether, sometimes obtained as the main product.

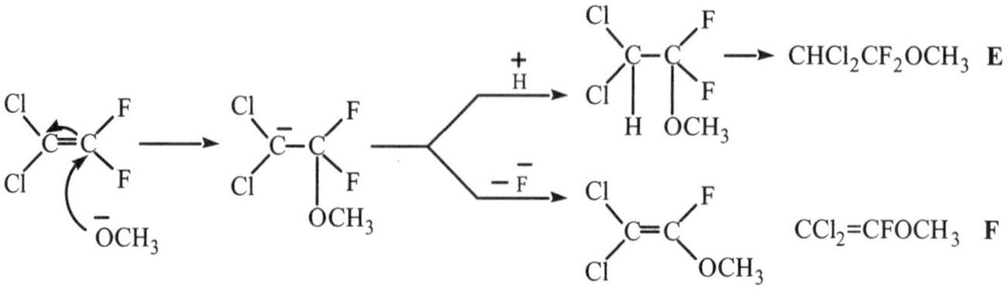

Usually, both products are formed, and their ratio depends on the nature of the alcohol and on experimental conditions. The addition of alcohols takes place only on such haloolefins, which have at least one fluorine atom at the double bond [87].

➡️ EXPLANATION 63

As a rule, in nucleophilic additions to halogenated alkenes, the nucleophile always becomes attached to that carbon of the double bond that has lower electron density. In this case, back-donation of electrons by fluorine is stronger in the difluoromethylene group, and nitrogen therefore attacks the difluoromethylene group while proton adds to the other carbon. Because of the basicity of the reaction mixture, hydrogen fluoride is eliminated, and an imide fluoride, 3-chloro-2,3-difluoro-1-phenyl-1-azapropene is formed:

$$CClF{=}CF_2 \longleftrightarrow \overset{\delta-}{CClF}{=}\overset{\delta+}{CF_2} \longleftrightarrow \overset{-}{CClF}{=}\overset{+}{CF_2} \xrightarrow{C_6H_5NH_2} \overset{H-\!\!-NHC_6H_5}{CClF{=}CF_2}$$

$$\xrightarrow{-\ HF} CHClFCF{=}NC_6H_5$$

This product undergoes addition of another molecule of aniline. The addition product loses proton and fluoride anion and affords **G**, *N,N'*-diphenylamidine of chlorofluoroacetic acid [88].

$$\overset{\delta+}{CHClFCF}{=}\overset{\delta-}{NC_6H_5} + H_2NC_6H_5 \longrightarrow \underset{NHC_6H_5}{CHClFCF{-}NHC_6H_5} \xrightarrow{-HF} \underset{NC_6H_5}{CHClFC{-}NHC_6H_5}$$

➡️ EXPLANATION 64

Addition of dimethylamine to chlorotrifluoroethylene occurs in such a way that the nucleophile, dimethylamine, joins the carbon having two fluorine atoms, because this carbon has lower electron density owing to strong electron back-donation by fluorine.

$$\overset{\delta-}{CClF}{=}\overset{\delta+}{CF_2} \longleftrightarrow \left(\overset{-}{CClF}{-}\overset{+}{CF_2} \longrightarrow CHClFCF_2N(CH_3)_2 \atop H{-\!\!-}N(CH_3)_2 \right)$$

From the initial addition product, hydrogen fluoride is ejected in the basic medium, and another molecule of dimethylamine is added across the double bond.

$$CHClF—CF_2N(CH_3)_2 \xrightarrow{-HF} CClF{=}CFN(CH_3)_2 \longrightarrow$$
$$H\text{——}N(CH_3)_2$$

+

$$CHClF—CF\overset{N(CH_3)_2}{\underset{N(CH_3)_2}{}} \xrightarrow{-HF} CClF{=}C\overset{N(CH_3)_2}{\underset{N(CH_3)_2}{}}$$

The same process of addition of dimethylamine followed by elimination of hydrogen fluoride, and once hydrogen chloride, repeats itself until **H**, a fluorine-free product, tetrakis(dimethylamino)ethylene, is obtained [88].

$$\overset{\delta+}{C}ClF{=}\overset{\delta-}{C}\overset{N(CH_3)_2}{\underset{N(CH_3)_2}{}} \longrightarrow (CH_3)_2NCClF—CH\overset{N(CH_3)_2}{\underset{N(CH_3)_2}{}} \xrightarrow{-HCl} (CH_3)_2N\overset{\delta+}{C}\overset{\delta-}{F}{=}C\overset{N(CH_3)_2}{\underset{N(CH_3)_2}{}}$$
$$(CH_3)_2N\text{——}H$$

$$\xrightarrow{(CH_3)_2NH} (CH_3)_2NCF{=}C\overset{(CH_3)_2N\text{——}H\ \ N(CH_3)_2}{\underset{N(CH_3)_2}{}} \longrightarrow \overset{(CH_3)_2N}{\underset{(CH_3)_2N}{}}CF—CH\overset{N(CH_3)_2}{\underset{N(CH_3)_2}{}}$$

$$\xrightarrow{-HF} \overset{(CH_3)_2N}{\underset{(CH_3)_2N}{}}C{=}C\overset{N(CH_3)_2}{\underset{N(CH_3)_2}{}} \quad \mathbf{H}$$

⟳ EXPLANATION 65

In contrast to nucleophilic additions of alcohols to fluorinated alkenes, where the reagent is alkoxide anion, in free-radical addition of alcohols the reacting species is a free radical generated by homolysis of the bond between hydrogen and the alkyl group.

$$CH_3OH + (C_6H_5COO)_2 \longrightarrow C_6H_5COOH + {}^{\bullet}CH_2OH + C_6H_5COO^{\bullet}$$

The free radical adds to the fluorinated alkene in such a way as to create a more stable free-radical intermediate. In perfluoalkene such as 1-perfluorobutene, the hydroxymethylene radical joins the sterically more accessible end of the double bond.

$$CF_3CF_2CF=CF_2 + \cdot CH_2OH \longrightarrow CF_3CF_2\overset{\bullet}{C}FCF_2CH_2OH$$

This intermediate free radical abstracts hydrogen from the methyl group of methanol, and thus propagates a chain process leading to the hydroxymethyl derivative, 1H, 1H, 3H-perfluoropentanol **I** [89]. Similar addition of alcohols to fluorinated alkenes occurs under γ-irradiation.

$$CF_3CF_2\overset{\bullet}{C}FCF_2CH_2OH + CH_3OH \longrightarrow CF_3CF_2CHFCF_2CH_2OH + \cdot CH_2OH$$

$$\mathbf{I}$$

⟱ EXPLANATION 66

The electron-deficient double bond in 1H-pentafluoropropene reacts easily with nucleophiles such as azide anion to form primarily an addition product 1-azido-1,2,3,3,3-pentafluoropropane. In the basic medium, unsaturated 1-azido-2,3,3,3-tetrafluoropropene is also formed, either by elimination of hydrogen fluoride from the addition product, or by direct displacement of fluorine by S_N2 mechanism [90].

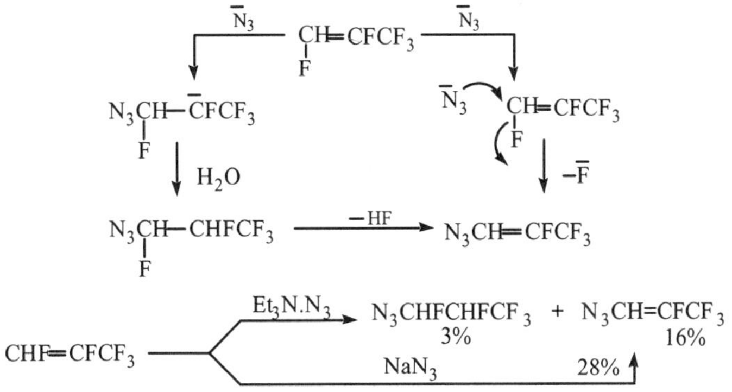

⟸ EXPLANATION 67

Aromatic fluorine is replaced by nucleophiles only if the electron density of the aromatic ring is lowered by the presence of electron-withdrawing substituents such as nitro groups. In this case, inductive effect of the four chlorine atoms depletes the electron density of the benzene ring. Inductive effect of fluorine atoms further decreases the electron density at the carbons to which they are attached. The lowest electron density is therefore at carbons 1 and 2.

The nucleophiles, methoxide anions, attack the carbon atoms linked to fluorine, and thus a nonaromatic "Meisenheimer complex" [91] is formed in the rate-determining step of the reaction. Rearomatization is a strong driving force for ejecting the fluoride leaving groups in a subsequent fast step [92] and forming compound **J**.

Low electron density of the aromatic ring is also achieved by complexing with transition metals. In chromium tricarbonyl fluorobenzene, the fluorine is replaced by a cyano group under very mild conditions [93].

In the case of 2,4-dinitrohalobenzenes, nitro groups decrease electron density of the benzene ring, especially in *ortho* and *para* positions to the halogen.

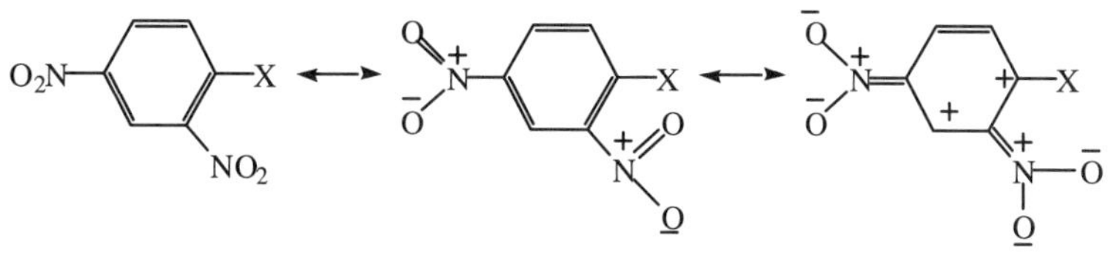

The electron density at the carbon holding the halogen, moreover, is decreased by the inductive effect of the halogen.

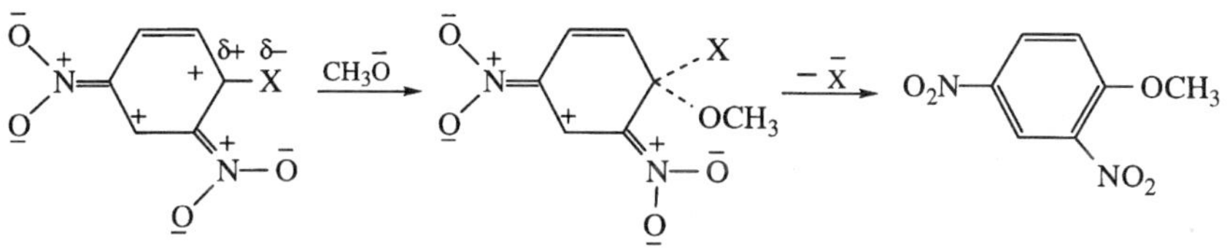

Thus, carbon 1 of 2,4-dinitrohalobenzene has the lowest electron density, and the halogen in position 1 is easily displaced by a nuclephile in a rate-determining step of an S_NAr reaction with hydroxide, alkoxide, and primary or secondary amino compounds [94]. As the nucleophile becomes attached to the carbon, the compound is converted to an unstable nonaromatic species, the "Meisenheimer complex" [91]. Restoration of aromaticity facilitates elimination of the halogen in a fast step, in this particular case giving 2,4-dinitroanisole.

The rate of displacement of a halogen by a nucleophile depends on the inductive electron withdrawal at the carbon to which the halogen is bonded. The inductive effect of fluorine is much stronger than that of chlorine. Consequently, the electron density at the carbon holding fluorine is lower than at that holding chlorine. Although fluoride ion is a worse leaving group than chloride ion, the carbon-halogen bond strength is unimportant, because the breaking of the bond occurs in a fast, non-rate-determining step. The relative reactivity of halogen toward nucleophiles is shown in the table below, which shows the difference (relative to $Cl = 1$) in the reactivity of compounds having one or two activating groups [95, 96].

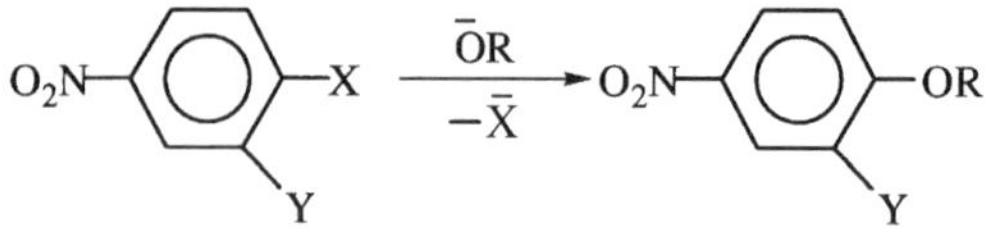

X=	F	Cl	Br	I		
Y=H	228	1	0.87	0.074	R=C$_2$H$_5$	[95]
Y=NO$_2$	581	1			R=CH$_3$	[96]

⟶ EXPLANATION 69

Nucleophilic displacement of halogens in aromatic rings by S$_N$Ar reaction could involve, in the case of fluoropentachlorobenzene, either chlorine or fluorine. Breaking the carbon-chlorine bond is easier than breaking the carbon-fluorine bond. However, the rate-determining step of the displacement of the halogen is nucleophilic attack at the aromatic carbon atom. This occurs at the place of the lowest electron density, that is, at the carbon linked to fluorine. After the formation of the nonaromatic intermediate, the tendency to rearomatize is so strong that the energy necessary for breaking of the carbon-halogen bond is of little importance. The product of the reaction results from the replacement of fluorine and is compound **K**, *p*-methoxyphenoxypentachlorobenzene [97].

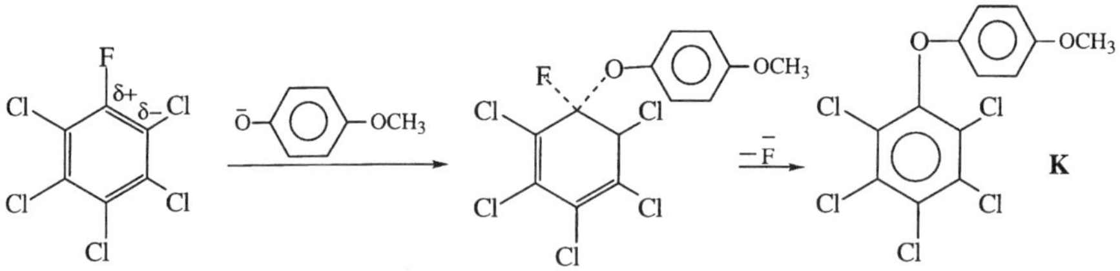

⟶ EXPLANATION 70

Explanation 70 is a big surprise. On heating for weeks at 90°C with a large excess of sodium salt of a hydroxy compound in *N,N'*-dimethyl-imidazolid-2-one, all fluorine atoms are replaced by the residues of the hydroxy compounds such as *m*-cresol, 2-naphthol, or 2-hydroxytetralin. Compound **L** is octakis(alkoxy)- or octakis(aryloxy)naphthalene [98].

R = *m*-cresyl, tetralyl, 2-naphthyl L =

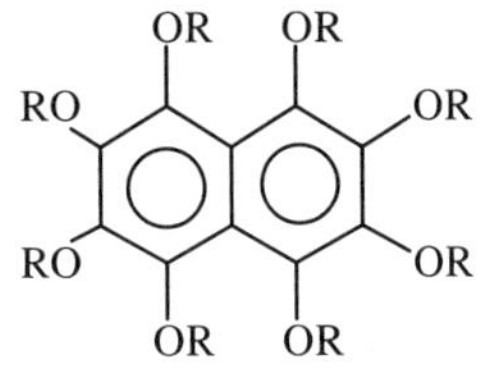

⟢ EXPLANATION 71

In the presence of benzophenone as a sensitizer, ultraviolet irradiation of methanol generates free radicals •CH$_2$OH. These radicals react with perfluoropyridine in electron-deficient positions such as α- and γ-positions. Elimination of hydrogen fluoride gives 4-hydroxymethylperfluoropyridine, compound **M** [99].

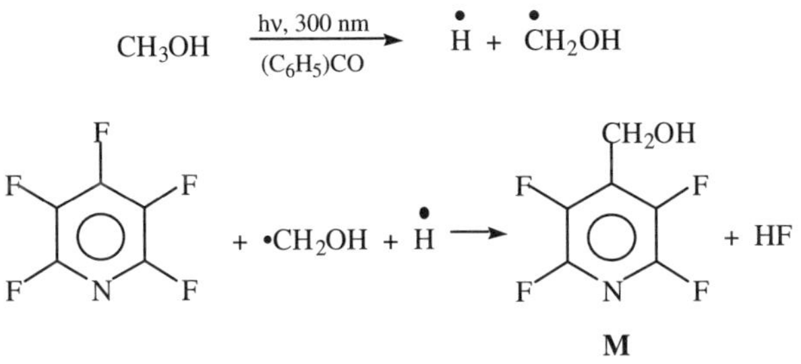

⟢ EXPLANATION 72

Both positions, *ortho-* and *para-* to the nitro group, are activated for the replacement by a nucleophile, in this case a sulfide anion. The nucleophilic attack takes place at the position of the lowest electron density. Since the inductive effect of fluorine is much stronger than that of bromine, the electron density is lower in position 4 than in position 2. Therefore, the sulfide anion displaces fluorine and not bromine [100].

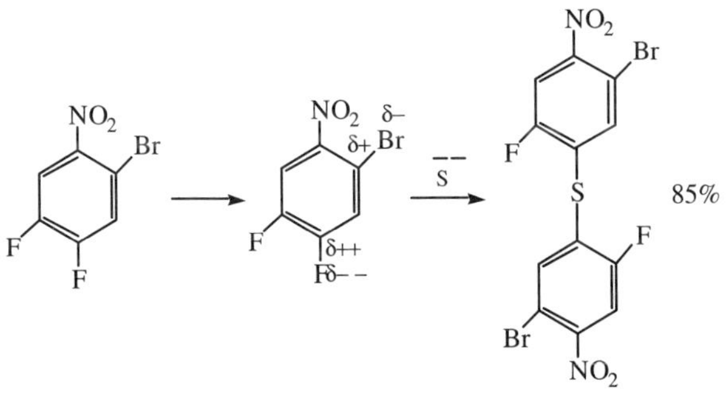

⟢ EXPLANATION 73

In mixed trifluoroacetic anhydrides, the strong electron-withdrawing tri-fluoroacetyl group polarizes the bonds to the anhydride oxygen in such a way that it tends to split to acylium cation and trifluoroacetate anion [101]. Generally, the products of the reaction of hydroxy compounds with acyl trifluoroacetates are esters of carboxylic acids and trifluoroacetic acid, because the acyl carbonyl is better fit for a nucleophilic attack than the carbonyl next to the trifluoromethyl group [101].

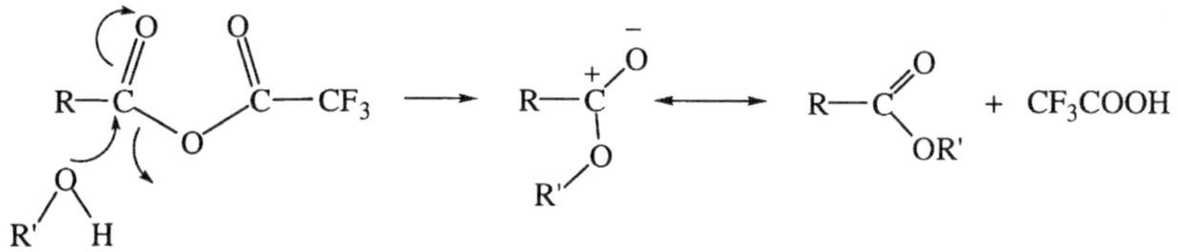

However, the products may also be trifluoroacetates of the hydroxy compounds. Trifluoroacetates are favored with hydroxy compounds such as nonpolar alcohols, which are better nucleophiles than the more acidic alcohols containing polar groups such as trifluoromethyl, or phenols whose acidity is higher than that of nonpolar alcohols by as much as six or seven orders of magnitude. In addition, formation of trifluoroacetates is enhanced by addition of carbon tetrachloride, trifluoroacetic acid, or both. The presence of trifluoracetic acid may affect the reaction equilibrium between the trifluoroacetate of the hydroxy compound and the carboxylic acid by reacting with the byproduct of the reaction, carboxylic acid, and thus increase the formation of the trifluoroacetate (Guldberg-Waage's law).

$$RCOOCOCF_3 + R'OH \rightleftharpoons \begin{array}{l} RCOOR' + CF_3COOH \\ + \\ RCOOH + CF_3COOR' \end{array}$$

The following table shows results of treatment of hydroxy compounds with benzoyl trifluoroacetate (1.3 mmol) at 20°C for 24 hours [102].

Alcohol	1.3 mol CF$_3$CO$_2$H		0.5 ml CCl$_4$				CF$_3$CO$_2$H + CCl$_4$	
R'OH	R'OCO-C$_6$H$_5$	R'OCO-CF$_3$	R'OCO-C$_6$H$_5$	R'OCO-CF$_3$	R'OCO-C$_6$H$_5$	R'OCO-CF$_3$	R'OCO-OC$_6$H$_5$	R'OCO-CF$_3$
CH$_3$CH$_2$OH	12%	84%	13%	81%		96%		97%
CF$_3$CH$_2$OH	44%	53%	82%	17%	20%	67%	80%	19%
C$_6$H$_5$HOH	75%	25%	88%	11%	76%	19%	93%	7%

13 Aldol-Type Condensations

⟹ **EXPLANATION 74**

The probable mechanism of the reaction is first addition of piperidine to formaldehyde to form *N*-hydroxymethylpiperidine. This reacts with acidic hydrogens of the methyl group of 1,1,1-trifluoroacetone, which is hydrated because the trifluoromethyl group next to carbonyl stabilizes the hydrated form of the carbonyl compound. The resulting dihydrate is further stabilized by hydrogen bonds to the piperidine nitrogen [103].

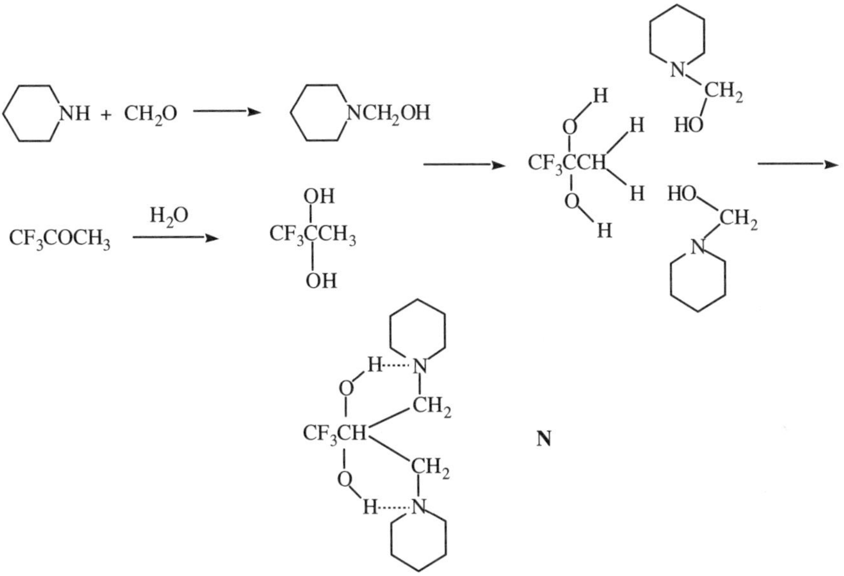

⟹ **EXPLANATION 75**

Hydrogen atoms on the carbon linked to phosphorus are acidic enough to protonate the oxygen atom of the carbonyl group while the rest of the phosphine adds to the carbonyl carbon. Owing to the strong affinity of phosphorus to oxygen, the intermediate splits into dibutylhydroxyphosphine (dibutyloxophosphorane) and a mixture of *E*- and *Z*- 2-phenyl-1,1,1-trifluorohex-2-ene (compounds **O** and **P**, respectively) [*104*].

O **P**

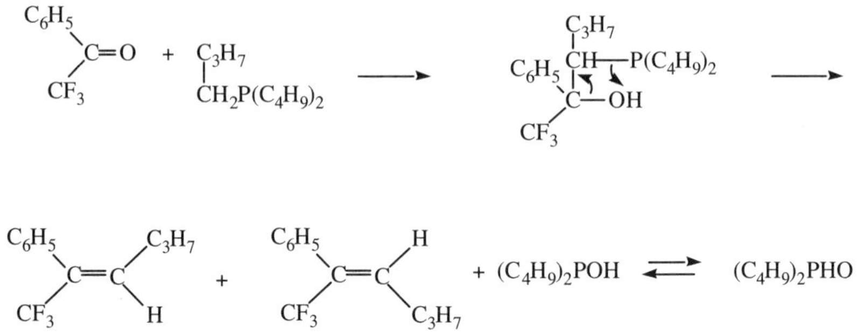

EXPLANATION 76

First, dibromodifluoromethane and hexamethylphosporous amide form a phosphonium salt. The intermediate thus formed reacts with tris(dimethylamino)phosphine to give dibromotris(dimethylamino)phosphorane and an ylide with difluoromethylene group attached to phosphorus, difluoromethylenetris(dimethylamino)phosphorane.

$$CBr_2F_2 + [(CH_3)_2N]_3P \longrightarrow \{[(CH_3)_2N]_3\overset{+}{P}CBrF_2\}\,\overset{-}{Br} \xrightarrow{[(CH_3)_2N]_3P}$$

$$[(CH_3)_2N]_3PBr_2 + [(CH_3)_2N]_3P{=}CF_2 \longleftrightarrow [(CH_3)_2N]_3\overset{+}{P}\text{-}\overset{-}{C}F_2$$

The ylide reacts with the ketone in the sense of the Wittig reaction in that the difluoromethylene group is attached to the carbonyl carbon while the oxygen forms tris(dimethylamino)phosphine oxide. Another mechanism has been suggested by the author of this reaction [*105*].

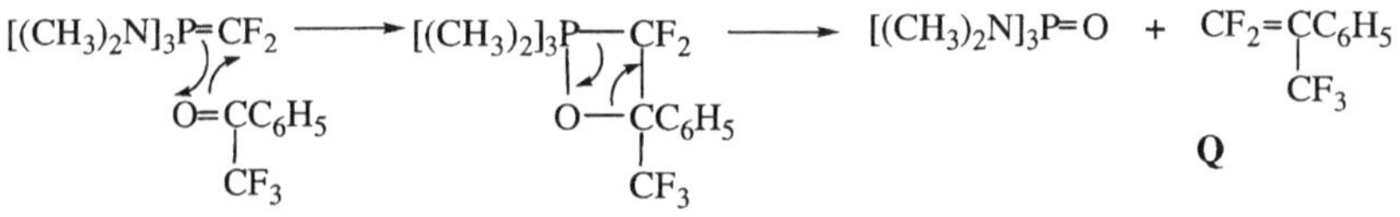

This and similar reactions are suitable for the synthesis of fluoro alkenes with terminal fluorine-containing groups. The final product of this reaction is **Q**, 2-phenyl-1,1,3,3,3-pentafluoropropene [*105*].

EXPLANATION 77

The phosphorus ylide adds to the triple bond of trifluoroacetonitrile to form a four-membered cyclic intermediate that is further transformed to another ylide with a phosphorus-nitrogen double bond **R.** This product is then cleaved with water to form **S**, a β-keto ester, ethyl 4,4,4-trifluoroac-etoacetate [*106*].

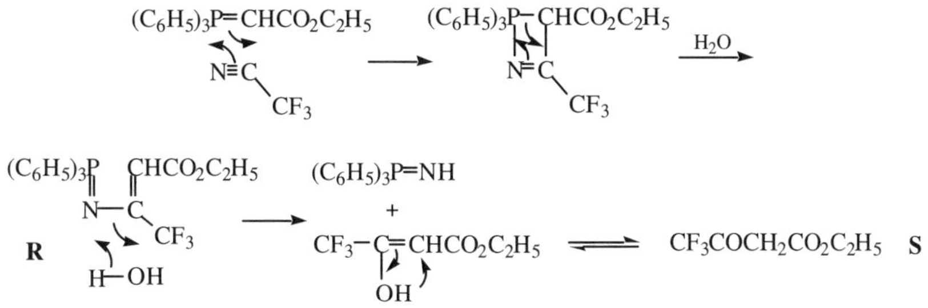

EXPLANATION 78

The strong inductive effect of the three trifluoromethyl groups polarizes the bond between the carbon and hydrogen and makes the hydrogen very acidic (pKa = 7). In basic media, the compound is deprotonated, and the tris(trifluoromethyl)carbanion adds to the double bond of acrylonitrile in the sense of the Michael addition. It joins the β-position of acrylonitrile because of its lower electron density. The product is 4,4-bis(trifluoro-methyl)-5,5,5-trifluorovaleronitrile **T** [*107*].

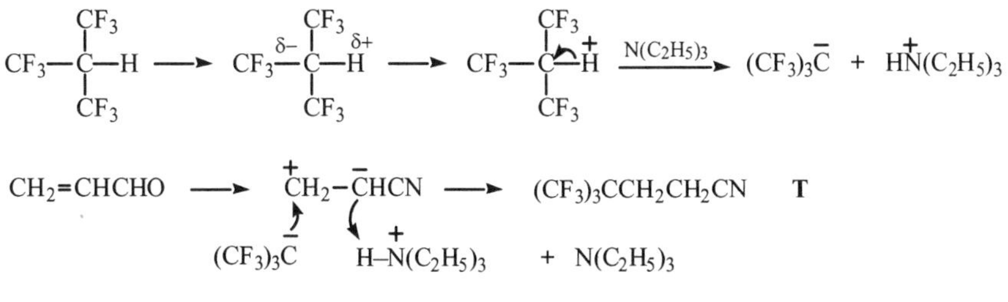

---

14 Organometallic Syntheses

EXPLANATION 79

Reaction of ethyl chlorofluoroacetate with phenylmagnesium bromide at −50° affords **U**, a ketone, chlorofluoromethyl phenyl ketone. Treatment

of the ketone with methylmagnesium bromide gives the expected intermediate **V**, 2-bromomagnesiumoxy-1-chloro-1-fluoro-2-phenylpropane.

$$CHClFC\overset{O}{\underset{OC_2H_5}{\diagdown}} + C_6H_5MgBr \longrightarrow CHClFC\overset{OMgBr}{\underset{C_6H_5}{\diagdown}}OC_2H_5 \xrightarrow{-C_2H_5OMgBr} CHClFCOC_6H_5 \quad U$$

$$CHClF\underset{C_6H_5}{\overset{|}{C}}=O + CH_3MgBr \longrightarrow CHClF\overset{CH_3}{\underset{OMgBr}{\overset{|}{C}}}C_6H_5 \quad V$$

$$\overset{Cl}{\underset{F}{\diagup}}\underset{BrMgO}{\overset{CH_3}{|}}CHCC_6H_5 \xrightarrow{-MgBrF} \overset{Cl}{\underset{}{}}\overset{CH_3}{CH-CC_6H_5} \longrightarrow OHCCClC_6H_5 \quad W$$

Compound **V** eliminates magnesium bromofluoride and is probably
converted to an epoxide, which rearranges to compound **W**, 2-chloro-2-
phenylpropanal. Elimination of magnesium bromofluoride instead of
magnesium bromochloride is probably due to stronger affinity of magnesium to fluoride ion [*108*].

➭ EXPLANATION 80

The resonance effect of fluorine atoms at the double bond, and the inductive effect of fluorines in the allylic position polarize the double bond so
that carbon 1 is most electrophilic. The phenyl carbanion joins carbon 1
to form the primary reaction product **X** [*109*].

$$C_6H_5MgBr + \overset{\delta+}{C}F_2=\overset{\delta-}{C}FCF_2Cl \longrightarrow [C_6H_5CF_2\overset{-}{C}FCF_2Cl]\overset{+}{M}gBr \quad X$$

From the organometallic intermediate **X**, either chloride or fluoride
anion is ejected, and thus **Y**, 3-phenylpentafluoropropene, or **Z**, 3-chloro-
1-phenyltetrafluoropropene, is formed.

$$[C_6H_5CF_2\overset{-}{C}FCF_2Cl]\overset{+}{M}gBr$$

$$\xleftarrow[\underset{-\overset{+}{M}gBr}{-\overset{-}{Cl}}]{} \qquad \xrightarrow[\underset{-\overset{+}{M}gBr}{-\overset{-}{F}}]{}$$

$$C_6H_5CF_2\overset{-}{C}F\text{-}\overset{+}{C}F_2 \qquad\qquad C_6H_5\overset{+}{C}F\text{-}\overset{-}{C}FCF_2Cl$$

$$C_6H_5CF_2CF=CF_2 \quad Y \qquad\qquad Z \quad C_6H_5CF=CFCF_2Cl$$

Although fluoride anion is a worse leaving group than chloride anion, the product formed by ejection of fluoride, **Z**, predominates, probably because its elimination creates a double bond conjugated with the aromatic ring.

⟵ EXPLANATION 81

Difluoromethylene groups flanking the double bond decrease the electron density of the bond and thus activate it for nucleophilic reactions. The nucleophile here is ethyl carbanion, which adds to the double bond. Chlorine atom, linked to the same carbon, is eliminated, and thus 1-chloro-2-ethylhexafluorocyclopentene is formed.

The same reaction is repeated at the neighboring carbon of the double bond and gives **A**, 1,2-diethylhexafluorocyclopentene:

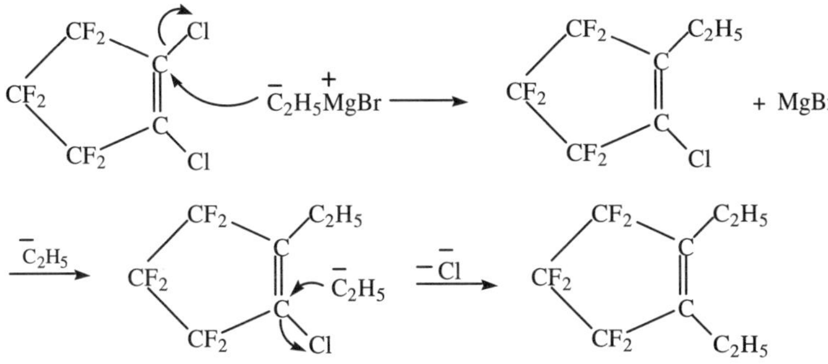

But there is still another possibility of a reaction of the chlorine-containing unsaturated product, 1-chloro-2-ethylhexafluorocyclopentene. Its double bond is polarized by the inductive effect of chlorine in such a way that higher electron density is at the carbon linked to chlorine. The positive charge on the carbon bonded to the ethyl group causes the ethyl carbanion to join this carbon, and the carbanion thus formed ejects fluoride ion from the neighboring difluoromethylene group and generates the double bond to give **B**, 2-chloro-3,3-diethylpentafluorocyclopentene [*110*].

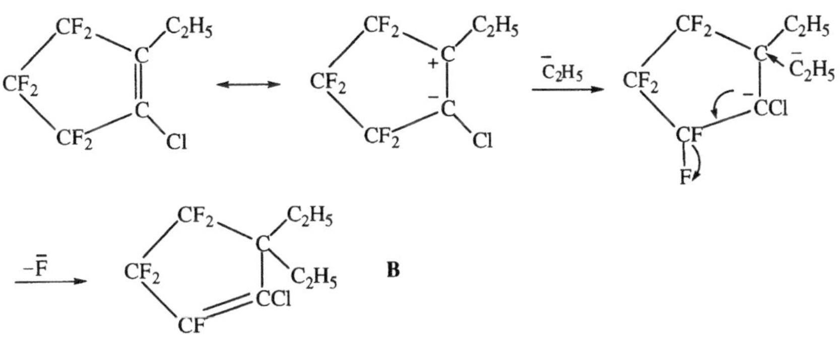

EXPLANATION 82

Both hydrogen and chlorine may be displaced by lithium. In this case, hydrogen is displaced preferentially to give 1,2-dichloro-2-fluorovinyllithium. The organometallic compound is then treated with acetone, and the intermediate is decomposed with water to afford 3,4-dichloro-4-fluoro-2-methyl-3-buten-2-ol, 1,2-dichloro-2-fluorovinyldimethylcarbinol; thus, X = Cl [111].

$$CClF{=}CHCl \xrightarrow{C_4H_9Li} CClF{=}CLiCl \xrightarrow{(CH_3)_2CO} CClF{=}CClC\underset{CH_3}{\overset{CH_3}{|}}-OLi$$

$$\xrightarrow[-LiOH]{H_2O} CClF{=}CClC\underset{CH_3}{\overset{CH_3}{|}}OH \quad X = Cl$$

EXPLANATION 83

Trifluoroethylene reacts with butyllithium to give **C**, trifluorovinyllithium. This reacts with acetone in the usual way to give, after hydrolysis, **D**, dimethyltrifluorovinylcarbinol, 2-methyl-3,4,4-trifluoro-3-buten-2-ol.

$$CF_2{=}CHF \xrightarrow{C_4H_9Li} CF_2{=}CFLi \xrightarrow{(CH_3)_2CO} CF_2{=}CFC\underset{CH_3}{\overset{CH_3}{|}}OLi \xrightarrow[-LiOH]{H_2O} CF_2{=}CFC\underset{CH_3}{\overset{CH_3}{|}}OH \quad D$$

$$\phantom{CF_2{=}CHF \xrightarrow{C_4H_9Li}} \mathbf{C}$$

The double bond is easily hydrated. Elimination of hydrogen fluoride yields an acyl fluoride which is ultimately hydrolyzed to **E**, 2-fluoro-3-methylbuten-2-oic acid [111].

$$CF_2{=}CFC\underset{CH_3}{\overset{CH_3}{|}}OH \xrightarrow{H_2O} F{-}CF{-}CF{-}C\underset{H}{\overset{CH_3}{|}}CH_3 \xrightarrow[-H_2O]{-HF} FCO{-}CF{=}C\overset{CH_3}{\underset{CH_3}{<}} \xrightarrow[-HF]{H_2O} (CH_3)_2C{=}CFCO_2H$$

$$\mathbf{E}$$

EXPLANATION 84

Butyllithium replaces with lithium both the hydrogen in 1-chloro-1,2-difluoroethylene and the chlorine in another molecule of the same compound, in a ratio of 1:2. Subsequent treatment of the reaction mixture with carbon dioxide, water, and esterification with methanol gives a mixture of methyl 3-chloro-2,3-difluoroacrylate (**F**) and methyl 2,3-difluoro-acrylate (**G**) in a ratio 1:2 [*112*].

$$CClF{=}CHF \xrightarrow{C_4H_9Li} CClF{=}CFLi \quad + \quad CHF{=}CFLi$$

$$\downarrow {\scriptstyle CO_2,\ H_2O} \qquad\qquad\quad \downarrow {\scriptstyle CO_2,\ H_2O}$$

$$CClF{=}CFCO_2H \qquad\quad CHF{=}CFCO_2H$$

$$\downarrow {\scriptstyle CH_3OH} \qquad\qquad\qquad \downarrow {\scriptstyle CH_3OH}$$

$$\mathbf{F} \quad CClF{=}CFCO_2CH_3 \qquad CHF{=}CFCO_2CH_3 \quad \mathbf{G}$$

⇨ EXPLANATION 85

1,1,1-Trichlorotrifluoroethane and zinc in diethyl ether or dimethylfor-mamide (DMF) form an organometallic complex with diethyl ether [*113*] or dimethylformamide [*114*]. This complex reacts with aldehydes similarly as the Grignard reagents to form primarily the expected alco-hols. In the presence of acetic anhydride, the alcohol is converted to the corresponding olefin, predominantly with *Z* configuration. In the case of benzaldehyde, *Z*- and *E*- 2-chloro-1-phenyl-3,3,3-trifluoropropene are the products **H** and **I** [*114*].

$$CF_3CCl_3 \ + \ {>}2\ Zn \xrightarrow{(C_2H_5)_2O} CF_3CCl_2ZnCl{\cdot}\ 2(C_2H_5)_2O$$

$$CF_3CCl_3 \ + \ {>}2\ Zn \xrightarrow{HCON(CH_2)_2} CF_3CCl_2ZnCl{\cdot}\ 2HCON(CH_3)_2$$

$$C_6H_5CHO \ + \ CF_3CCl_2ZnCl \longrightarrow \underset{\underset{OZnCl}{|}}{C_6H_5CHCCl_2CF_3} \xrightarrow[\ 2.\ (CH_3CO)_2O\ ]{1.\ H_2O}$$

$$\underset{\underset{COCH_3}{\overset{|}{O}}\ Cl}{C_6H_5CH{-}CClCF_3} \longrightarrow CH_3COOCl \ + \ \begin{array}{c} \underset{H}{\overset{C_6H_5}{\diagdown}}C{=}C\underset{CF_3}{\overset{Cl}{\diagup}} \quad \mathbf{H} \\[4pt] \underset{H}{\overset{C_6H_5}{\diagdown}}C{=}C\underset{Cl}{\overset{CF_3}{\diagup}} \quad \mathbf{I} \end{array}$$

⇨ EXPLANATION 86

First, 1,1,1-trichlorotrifluoroethane reacts with zinc to form an organo-metallic complex with dimethylformamide (DMF):

$$CF_3CCl_3 + >2\ Zn \xrightarrow{DMF} CF_3CCl_2ZnCl \cdot 2\ HCON(CH_3)_2$$

The organozinc intermediate thus formed reacts with aldehydes as Grignard reagents do: to form alcohols. In the presence of aluminum chloride, elimination of chlorine and fluorine from the vicinal carbons of the dichlorotrifluoroethyl group generates halogenated allylic alkoxides that are protonated to allylic alcohols, in the present case **J**, 2-chloro-3,3-difluoro-1-phenylpropen-2-ol [*114*].

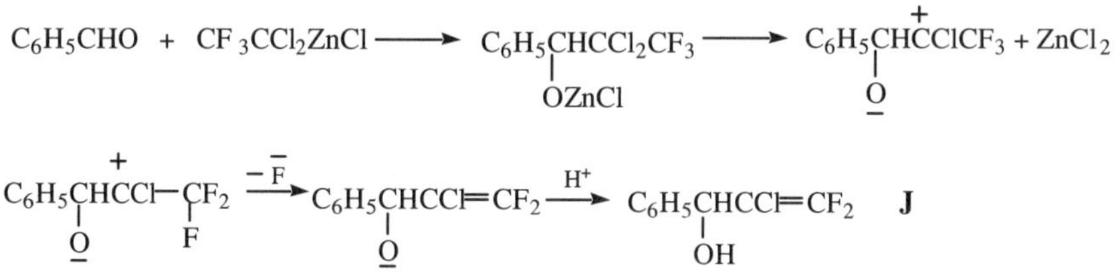

$$C_6H_5CHO + CF_3CCl_2ZnCl \longrightarrow \underset{\underset{OZnCl}{|}}{C_6H_5CHCCl_2CF_3} \longrightarrow \underset{\underset{O^-}{|}}{C_6H_5\overset{+}{C}HCClCF_3} + ZnCl_2$$

$$\underset{\underset{O^-}{|}}{C_6H_5\overset{+}{C}HCCl}\!-\!\underset{\underset{F}{|}}{CF_2} \xrightarrow{-\bar{F}} \underset{\underset{O^-}{|}}{C_6H_5CHCCl\!=\!CF_2} \xrightarrow{H^+} \underset{\underset{OH}{|}}{C_6H_5CHCCl\!=\!CF_2} \quad \mathbf{J}$$

⟵ EXPLANATION 87

First, an organometallic compound is formed from dibromodifluoromethane. Quite unexpectedly, trifluoromethyl copper is generated—who knows how?

$$CF_2Br_2 + Cu \longrightarrow CF_3Cu$$

Trifluoromethyl copper reacts with 4-chloro-3-nitrobenzotrifluoride in position 4. The carbanion of the organocopper compound, trifluoromethyl, attacks this position, because as a very strong nucleophile, it joins the carbon of the lowest electron density. The electron density of position 4 is decreased by the trifluoromethyl group in *para* position, and even more strongly by the nitro group in *ortho* position. The product is **K**, 3-nitro-1,4-bis-(trifluoromethyl)benzene [*115*].

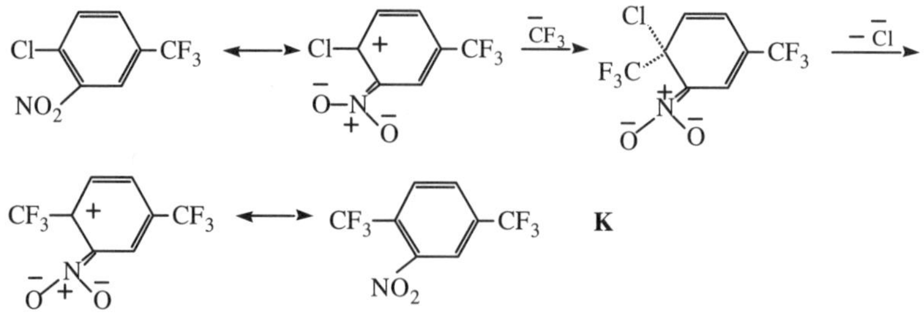

15   Additions

 EXPLANATION 88

The reaction in Explanation 88 is a nucleophilic addition of 2-amino-ethanol (ethanolamine) to an α,β-unsaturated ester. In basic media, proton can be abstracted from the hydroxy group as well as from the amino group of ethanolamine. Of the two anions thus formed, the anion with a free electron pair on nitrogen is more nucleophilic because hydrogen in the amino group is less acidic than hydrogen in the hydroxy group. Therefore, the attacking part of the ethanolamine molecule is the amino group.

To what carbon of the double bond will the nucleophile be attached? In α,β-unsaturated carbonyl compouds and nitriles, oxygen or nitrogen attracts electrons of the double bond and polarizes it so that the *beta* carbon has lower electron density. A clasical example is addition of water or ammonia to methyl acrylate:

$$CH_2{=}CH{-}C \overset{\delta-}{\underset{OCH_3}{\overset{O}{\Big\backslash}}} \longleftrightarrow CH_2{-}CH{-}C \overset{O}{\underset{OCH_3}{\Big\backslash}} \xrightarrow{NH_3} CH_2{-}CH{-}C \overset{O}{\underset{OCH_3}{\Big\backslash}}$$

$$\longrightarrow H_2NCH_2CH_2COOCH_3$$

In heavily fluorinated carbonyl compounds, a trifluoromethyl group also polarizes the double bond, but in the opposite way. In ethyl γ,γ,γ-trifluorocrotonate, polarization by the carbonyl function still predominates, and the nucleophile is bound to the *beta* carbon [*116*].

$$CF_3CH{=}CHCO_2C_2H_5 \xrightarrow[100°]{NH_3} CF_3CHCH_2CO_2C_2H_5 \quad 95\%$$
$$\underset{NH_2}{\big|}$$

However, if two trifluoromethyl groups are attached to the *beta* carbon, the cumulative effect of the two trifluoromethyl groups polarizes the double bond so that lower electron density is in *alpha* position, and the nucleophile becomes attached to the *alpha* carbon. The reaction product is **L**, ethyl 2-(β-hydroxyethylamino)-3-trifluoromethyl-4,4-trifluorobutanoate [*117*].

CF ... (chemical scheme)

$$CF_3 \underset{CF_3}{\overset{\delta-\ \ \delta+}{>}} C = CHCO_2C_2H_5 \longrightarrow CF_3 \underset{CF_3}{\overset{-}{>}} \overset{+}{C} - \overset{+}{C}HCO_2C_2H_5 \longrightarrow CF_3 \underset{CF_3}{>} CHCHCO_2C_2H_5 \quad \mathbf{L}$$

(with H–NH, CH₂CH₂OH, and NHCH₂CH₂OH groups)

EXPLANATION 89

The usual reaction of alkenes with conjugated dienes is the Diels-Alder synthesis. This addition does not apply, with a few exceptions, to cyclo-addition of fluorinated alkenes. The presence of fluorine atoms, especially if there are more of them around the double bond, retards the cycloadditions forming six-membered rings, and accelerates the (2 + 2) cycloaddition to form four-membered rings.

The bond strain in tetrafluoroethylene is much higher (41.2 kcal/mol) than that in ethylene (22.4 kcal). However, even though the strain in per-fluorocyclobutane (32.0 kcal/mol) is higher than that in butane (26.2 kcal/mol), the overall relief of bond strain in the ring closure to the four-membered ring is larger for tetrafluoroethylene than for ethylene (50.4 kcal/mol and 18.6 kcal/mol, respectively) [118].

Moreover, the carbon-fluorine bonds in tetrafluoroethylene are sp^3 hybridized, and the bond angle FCF in tetrafluoroethylene in the doubly bonded difluoromethylene group is 110° compared with the bond angle in HCH in ethylene (117°12′). The more fluorine atoms around the double bond in fluorinated alkenes, the easier it is to form four-membered rings.

The second problem in the addition of 1,1-dichlorodifluoroethylene to 1,3-butadiene is regiospecificity. Carbon 1 of 1,3-butadiene may become attached either to the carbon holding two chlorines or to the carbon with two fluorines. The cycloaddition of fluorinated alkenes is usually not a concerted four-center reaction in which the bonds are formed simultaneously or nearly so. Instead, the reaction is a stepwise biradical process in which the first step is formation of a free-radical intermediate with a single electron at that end of the double bond that can better accomodate it. That happens at the carbon linked to two chlorine atoms. Thus, a biradical is formed that cyclizes to form 2,2-dichloro-1,1-difluoro-3-vinylcyclobutane **M** [118, 119, 120].

$$CF_2=CCl_2 \qquad CF_2-CCl_2 \qquad CF_2-CCl_2 \qquad CF_2-CCl_2$$
$$CH_2=CH \qquad CH_2=CH \qquad CH_2-CH \qquad CH_2-CH$$
$$CH=CH_2 \qquad CH=CH_2 \qquad CH=CH_2 \qquad CH=CH_2$$
$$\mathbf{M}$$

At the temperature of 180–200°C, perfluoropropylene oxide ejects difluorocarbene and forms **N**, trifluoroacetyl fluoride (general method for the preparation of difluorocarbene). Difluorocarbene adds to 1,2-dichlorodifluoroethylene and yields **O**, 1,2-dichloroperfluorocyclopropane [121].

$$CF_3CF\overset{\diagup}{\underset{O}{\diagdown}}CF_2 \xrightarrow{185°} CF_3CF\overset{\diagup}{\underset{O}{\diagdown}}CF_2 \longrightarrow \underset{N}{CF_3CF} + \ :CF_2$$
$$\qquad\qquad\qquad\qquad\qquad\qquad\qquad \underset{O}{\parallel}$$

$$CClF{=}CClF + \ :CF_2 \longrightarrow CClF\overset{\diagup}{\underset{CF_2}{\diagdown}}CClF \quad \mathbf{O}$$

As in Explanation 89, these two compounds do not react in the sense of the Diels-Alder reaction. Instead, a cyclobutane derivative is formed by cycloaddition. The reaction is a free-radical process starting with single electrons on carbons 1 and 2 of perfluorovinylsulfur pentafluoride. Carbon 1 of the 1,3-butadiene joins the difluoromethylene group of the perfluorovinylsulfur pentafluoride because in this way, the newly formed biradical can better accommodate the single electron on the carbon next to the SF_5 group. The biradical closes the ring in such a way that two stereoisomers, **P** and **Q**, *cis*- and *trans*-2,3,3-trifluoro-1-vinyl-2-cyclobutylsulfur pentafluoride, are formed in equal amounts [122].

$$\begin{array}{c} CF_2{=}CFSF_5 \\ + \\ CH_2{=}CH{-}CH{=}CH_2 \end{array} \longrightarrow \begin{array}{c} \overset{\bullet}{CF_2}{-}\overset{\bullet}{CFSF_5} \\ \\ CH_2{=}CH{-}CH{=}CH_2 \end{array} \longrightarrow \begin{array}{c} CF_2{-}\overset{\bullet}{CFSF_5} \\ | \\ \overset{\bullet}{CH_2}{-}CH{-}CH{=}CH_2 \end{array}$$

$$\longrightarrow \begin{array}{c} CF_2{-}CF{-}SF_5 \\ | \qquad | \\ CH_2{-}CH{-}CH{=}CH_2 \end{array} \ \mathbf{P} \ + \ \begin{array}{c} CF_2{-}CF{-}SF_5 \\ | \qquad | \\ CH_2{-}CH\cdots CH{=}CH_2 \end{array} \ \mathbf{Q}$$

Free radicals with single electrons on the carbons of the trifluoroethylene combine with the double bond of butadiene in both possible ways: carbon of the difluoromethylene joins either the terminal carbon of the butadiene or the internal end of the double bond of the butadiene. The biradicals thus formed cyclize to **R**, 2,3,3-trifluoro-1-vinylcyclobutane, and **S**, 2,2,3-trifluoro-1-vinylcyclobutane, both as 1:1 mixtures of *cis*- and *trans*-isomers.

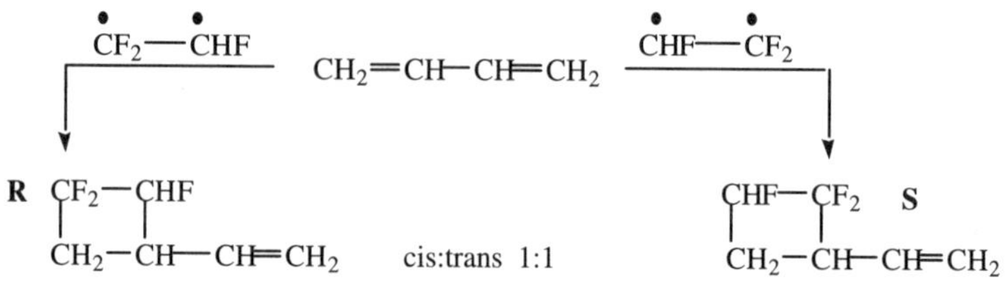

$$\overset{\bullet}{C}F_2-\overset{\bullet}{C}HF \qquad CH_2=CH-CH=CH_2 \qquad \overset{\bullet}{C}HF-\overset{\bullet}{C}F_2$$

$$\mathbf{R}\ \ \underset{CH_2-CH-CH=CH_2}{CF_2-CHF} \qquad \text{cis:trans 1:1} \qquad \underset{CH_2-CH-CH=CH_2}{CHF-CF_2}\ \ \mathbf{S}$$

In addition, a small amount of a six-membered ring compound is formed by the Diels-Alder reaction as a third product to give **T**, 4,4,5-tri-fluoro-1-cyclohexene. It is likely that both concerted and biradical mechanisms are involved [123].

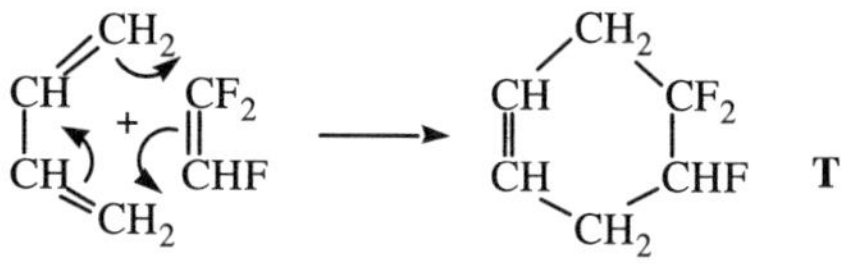

$$\underset{\underset{CH_2}{\overset{CH_2}{\Big|}}}{\overset{CH_2}{CH}}\ \ \underset{CHF}{\overset{CF_2}{\Big|}} \quad + \quad \longrightarrow \quad \underset{\underset{CH_2}{CH}}{\overset{CH_2}{CH}}\ \ \underset{CHF}{\overset{CF_2}{\Big|}}\quad \mathbf{T}$$

EXPLANATION 93

The reaction shown above is a bizarre example of a "crisscross" cyclo-addition. It starts as a (2 + 3) cycloaddition of one molecule of acetylene to positions 1 and 3 of the hexafluoroacetone azine to form an intermediate that by another (2 + 3) addition of another molecule of acetylene across the remaining nitrogen-carbon double bond forms another five-membered ring. In this way, a bicyclic system of two five-membered rings is formed to yield the final product [124, 125].

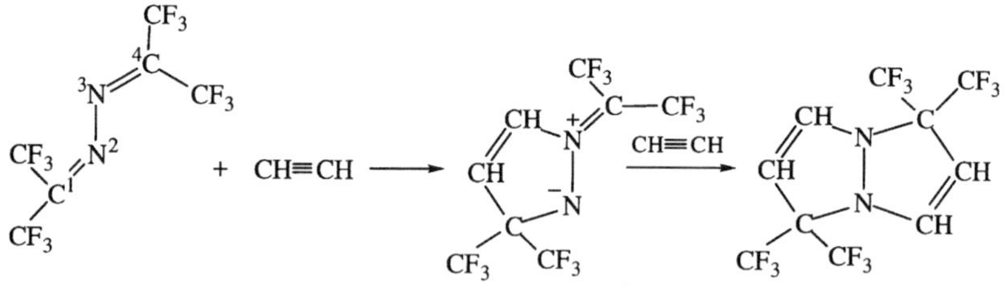

$$+ \quad CH\equiv CH \quad \longrightarrow \quad + \quad CH\equiv CH \quad \longrightarrow$$

EXPLANATION 94

The reaction of tris(*tert*-butyl)azete with trifluoroacetonitrile is a (4 + 2) cycloaddition followed by a *retro* (2 + 2) cycloaddition to afford 2-tri-fluoromethyl-4,5,6-tris(*tert*-butyl)pyrimidine [126].

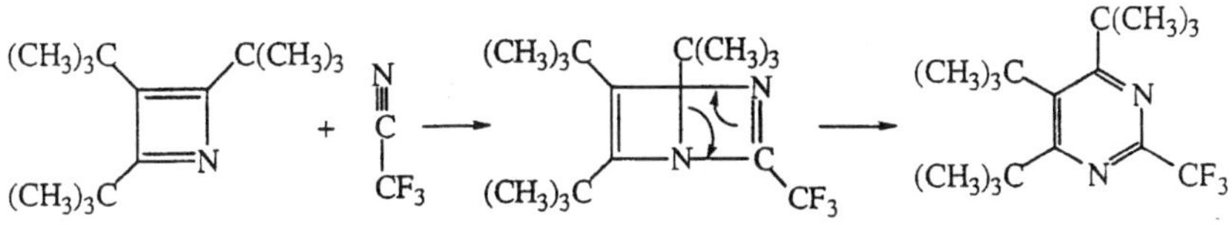

➥ EXPLANATION 95

Cycloadditions of fluorinated dienes are known to give cyclobutane rings. The original suggestion of the structure of the dimer of perfluoro-1,3-butadiene was perfluorotricyclo[4,2,0,0^{2,5}]octane [127]. The compound did not react with potassium permanganate. Fifteen years later, another formula was found correct based on physical methods. Instead of three four-membered rings, the product consists of three five-membered rings. Its name is perfluorotricyclo[3,3,0,0^{2,6}]octane [128].

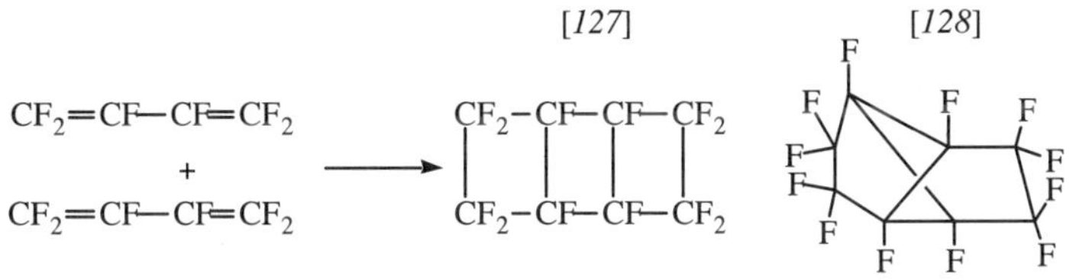

---

16 Eliminations

➥ EXPLANATION 96

In alkaline media, the cyclopropane ring is opened by a hydroxide anion in an S_N2 reaction. The anion attacks the least sterically hindered carbon of the ring, in this case the carbon having hydrogen attached to it. The bond to the carbon linked to the two methyl groups is broken, and from the negatively charged species, fluoride anion is ejected. The intermediate is dehydrated to *trans*-3-fluoro-4-methyl-1-phenylsulfonylpenta-1,3-diene [129].

$$C_6H_5O_2SCH_2CH\!-\!C\overset{CH_3}{\underset{CH_3}{<}} \quad \overset{\bar{}OH}{\longrightarrow} \quad C_6H_5O_2SCH_2CH\!-\!C\overset{CH_3}{\underset{CH_3}{<}} \quad \longrightarrow$$

(with CF$_2$ bridging groups and OH)

$$\underset{C_6H_5O_2SCH_2CH-\overset{|}{\underset{|}{C}}\!-\!\overset{F}{\underset{CH_3}{C}}<}{\overset{OH\quad F}{}} \quad \overset{-\bar{F}}{\longrightarrow} \quad C_6H_5O_2SCH_2CH\!-\!CF\!=\!C\overset{CH_3}{\underset{CH_3}{<}}$$

$$\overset{-H_2O}{\longrightarrow} \quad \underset{H}{\overset{C_6H_5O_2S}{>}}C\!=\!C\overset{H}{\underset{F}{<}}C\!=\!C\overset{CH_3}{\underset{CH_3}{<}}$$

⟵ EXPLANATION 97

Reactive halogens in *alpha* positions to carboxyl groups are easily replaced by nucleophiles, or eliminated as hydrogen halides. In the present case, elimination occurs easily because formation of a double bond conjugated with two carboxyl groups decreases the enthalpy of the reaction by a few kilocalories (resonance energy). Both hydrogen bromide or hydrogen fluoride can be eliminated. At first glance, elimination of hydrogen bromide would be expected because breaking the carbon-fluorine bond requires more energy than breaking the carbon-bromine bond.

However, there are several factors that facilitate elimination of hydrogen fluoride. Hydrogen on carbon adjacent to the carbon linked to fluorine is fairly acidic as it is α- to the carboxyl and β- to fluorine. Its elimination in basic media forms a carbanion in which the electron pair after deprotonation is stabilized by empty orbitals of fluorine. The same electron pair is stabilized also by bromine because bromine can accommodate the negative charge in its *d*-orbitals. In addition, in aqueous media, hydrogen bonding changes the nature of the leaving group, fluoride ion, to electrically neutral hydrogen fluoride, which is a much better leaving group than would be fluoride anion. Consequently, it is hydrogen fluoride that is split off via *cis*-elimination, probably by E1cb (carbanion) mechanism [*130*].

$$\underset{\overset{|}{\underset{CO_2C_2H_5}{}}}{\overset{CO_2C_2H_5}{\underset{F-\overset{|}{\underset{|}{C}}-H}{H-\overset{|}{\underset{|}{C}}-Br}}} \quad \overset{CH_3CO\bar{O}}{\underset{-CH_3COOH}{\longrightarrow}} \quad \underset{\overset{|}{\underset{CO_2C_2H_5}{}}}{\overset{CO_2C_2H_5}{\underset{F-\overset{|}{\underset{|}{C}}-H}{\bar{C}-Br}}} \quad \overset{H_2O}{\longrightarrow} \quad \underset{\overset{|}{\underset{CO_2C_2H_5}{}}}{\overset{CO_2C_2H_5}{\underset{F-\overset{|}{\underset{|}{C}}-H}{\bar{C}-Br}}} \quad \overset{-\bar{O}H}{\underset{-HF}{\longrightarrow}} \quad \underset{\overset{|}{\underset{CO_2C_2H_5}{}}}{\overset{CO_2C_2H_5}{\underset{CH}{CBr}}} \quad U$$

In the usual elimination of hydrogen halide from cyclic halides by the E2 mechanism, the dehydrohalogenation occurs only if the halogen and the vicinal hydrogens occupy antiperiplanar positions. In *cis*-1-bromo-2-fluorocyclohexane, there are two hydrogens that comply with this requirement: on carbon 2 and on carbon 6. As a result, only hydrogen bromide is eliminated to give **V**, 1-fluorocyclohexene, and **W**, 3-fluorocyclohexene. There are also two hydrogen atoms antiperiplanar to fluorine, but breaking of the carbon-fluorine bond requires much stronger bases, and under such conditions, hydrogen bromide is eliminated preferentially [*131, 132*].

In *trans*-1-bromo-2-fluorocyclohexane, E2 elimination occurs in the treatment of the compound with sodium methoxide or potassium *tert*-butoxide. There is only one hydrogen antiperiplanar to bromine, and its elimination leads to 3-fluorocyclohexene **W**. On the other hand, when, sodamide is used as a base, hydrogen fluoride is eliminated, not by the E2 mechanism but by a *cis*-elimination leading to **X**, 1-bromocyclohexene, probably by E1cb (carbanion) mechanism [*132*].

The clue to this rather exceptional reaction course is the possibility of the amide anion to form hydrogen bonds in the six-membered transition states. In the transition state of *trans*-1-bromo-2-fluorocyclohexane with both halogens in axial configurations, the most acidic hydrogen, *beta* to fluorine, forms a hydrogen bond to the nitrogen of the amide anion and is eliminated by combination with the amide anion. Fluorine, too, forms a hydrogen bond with the amide anion and is eliminated as hydrogen fluoride, which is a better leaving group than would be the fluoride anion. Such "double assistance" is not possible when alkoxides are used as bases. They can form hydrogen bonds to fluorine but are weaker bases than sodamide for elimination of *beta* hydrogen, which is undoubtedly the first reaction step. Similar double bonding occurs in the transition state with bromine and fluorine in equatorial configurations. The result is elimination of hydrogen fluoride to give 1-bromocyclohexene **X** [*132*].

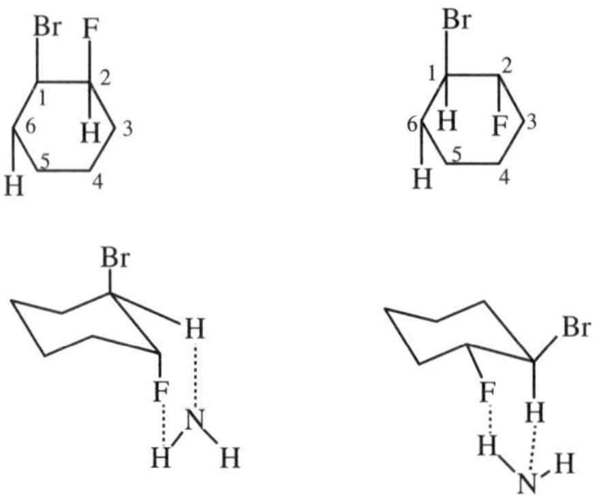

Br F
6 1 2 3
5 H
H 4

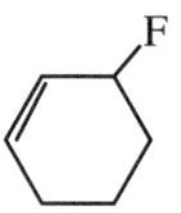

Br
1 6 H 2 F 3
5
H 4

Br
H
F
N
H H

Br
F
H
H
N H

$V = $ 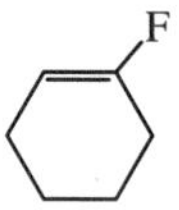 $W = $ $X = $

with F, F, Br labels on the cyclohexene rings respectively.

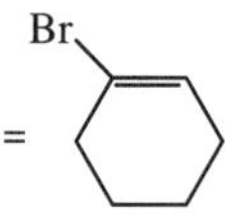

17 Rearrangements

EXPLANATION 99

The reaction of fluorohaloethanes with catalytic amounts of aluminum chloride or, better still, aluminum bromide, is peculiar to fluorohaloalkanes and alkenes. It is the migration of fluorine and other halogens to accumulate fluorine on one carbon [133].

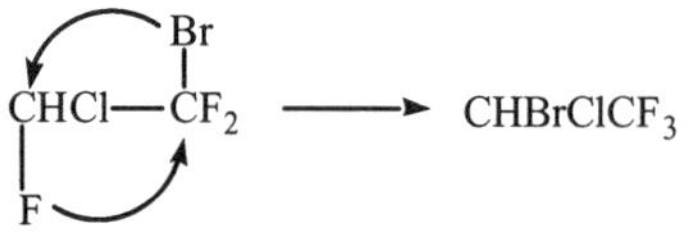

$$CHCl-CF_2 \xrightarrow{\quad Br \quad} CHBrClCF_3 \quad Y$$

The halogen exchange takes place intramolecularly [134]. Explanation of this reaction is difficult because it is unique in compounds containing halogens and at least two fluorine atoms. Maybe this rearrangement takes place because fluorine atoms like each other.

The above reaction is of particular interest because the product **Y** is the well-known inhalation anesthetic Halothane that replaced ether and other flammable and therefore dangerous anesthetics in the late 1950s. Nowadays, Halothane is, at least in the United States, replaced by even better fluorinated inhalation anesthetics.

Many other halofluoroethanes rearrange similarly and give products that are not always accessible by other reactions. A few examples are listed below to show the yields and reaction conditions [134, 135].

CCl_2FCCl_2F $\xrightarrow[50°]{AlCl_3}$ $CClF_2CCl_3$ 80-90% [134]

$CClF_2CCl_2F$ $\xrightarrow{AlCl_3, 50-55°, 5 h}$ CF_3CCl_3 90% [134]

$CBrF_2CBrClF$ $\xrightarrow[80-90°]{AlBr_3}$ CF_3CBr_2Cl 72-82% [135]

⟸ EXPLANATION 100

First, antimony trifluoride is treated with chlorine and thus antimony dichlorotrifluoride is formed. This compound catalyzes intramolecular rearrangement of fluorine with the simultaneous allylic shift of the double bond. As in Explanation 99, the tendency is to accumulate fluorine at the same carbon. The product is 1-chloroperfluoropropene **Z** [136].

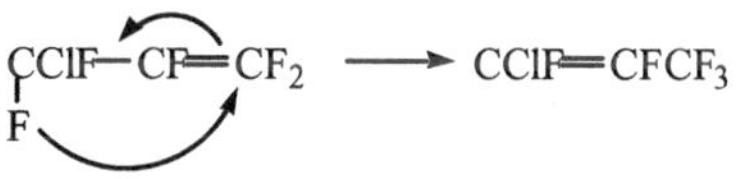

$$CClF{-}CF{=}CF_2 \longrightarrow CClF{=}CFCF_3 \quad \mathbf{Z}$$

⟸ EXPLANATION 101

Ammonia adds across the nitroso group to form an addition product, trifluoromethylaminohydroxylamine. This condenses with another molecule of trifluoronitrosomethane to form 3-hydroxy-1,3-bis(trifluoromethyl)-1-triazene, which decomposes to nitrogen and bis(trifluoromethyl)hydroxylamine **A**. Another mechanism of the reaction has also been proposed [137].

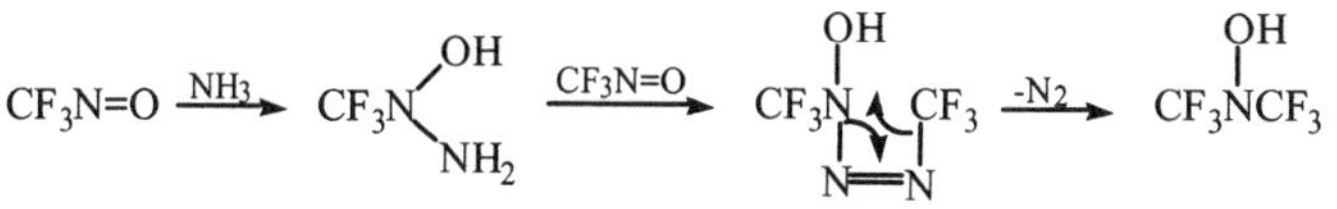

$$CF_3N{=}O \xrightarrow{NH_3} CF_3N\overset{OH}{\underset{NH_2}{\diagdown}} \xrightarrow{CF_3N{=}O} CF_3N\overset{OH}{\underset{N{=}N}{\diagdown}}CF_3 \xrightarrow{-N_2} CF_3\overset{OH}{N}CF_3 \quad \mathbf{A}$$

⟸ EXPLANATION 102

This somewhat unusual Hofmann degradation consists of several steps. In the first step, perfluorobutyramide is converted by silver oxide to its silver salt (**B**). This is then treated with bromine to give N-bromoperfluorobutyramide (**C**).

$$C_3F_7CONH_2 \xrightarrow{Ag_2O} \underset{\textbf{B}}{C_3F_7CONHAg} \xrightarrow{Br_2} \underset{\textbf{C}}{C_3F_7CONHBr}$$

Subsequent reaction with sodium hydroxide gives the sodium salt of *N*-bromoperfluorobutyramide. This intermediate when heated at 165–170°C affords the expected product of the Hofmann degradation, perfluoropropylisocyanate (**D**). More interesting is the product of the reaction of the sodium salt of *N*-bromoperfluorobutyramide with water: heptafluoropropyl bromide **E** [*138*].

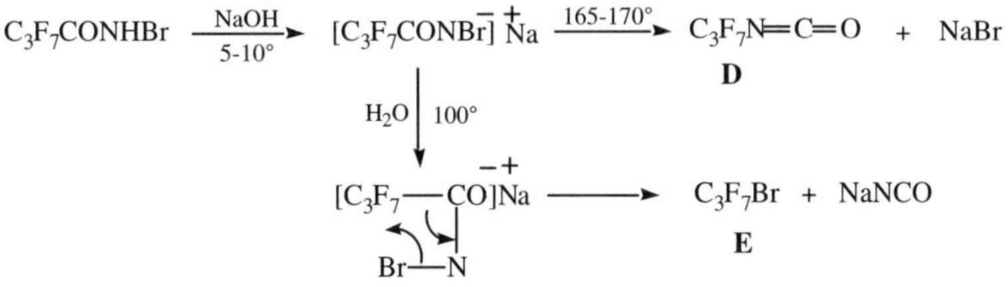

$$C_3F_7CONHBr \xrightarrow[5\text{-}10°]{NaOH} [C_3F_7CON\bar{Br}]\ \overset{+}{Na} \xrightarrow{165\text{-}170°} \underset{\textbf{D}}{C_3F_7N\text{=}C\text{=}O} \ + \ NaBr$$

$$\big\downarrow H_2O \quad 100°$$

$$[C_3F_7\text{—}\underset{}{CO}]\overset{-+}{Na} \longrightarrow \underset{\textbf{E}}{C_3F_7Br} \ + \ NaNCO$$

$$Br\text{—}N$$

⟹ EXPLANATION 103

Ultraviolet irradiation causes changes in carbon skeleton of many compounds. A classical example is conversion of hexafluorobenzene to hexafluoro Dewar benzene [*139*], and isomerization of benzene to three isomers. The same is true of hexakis(trifluoromethyl)benzene, which rearranges to the perfluorinated isomers of hexamethyl Dewar benzene (**F**), prismane (**G**), and benzvalene (**H**) [*140*].

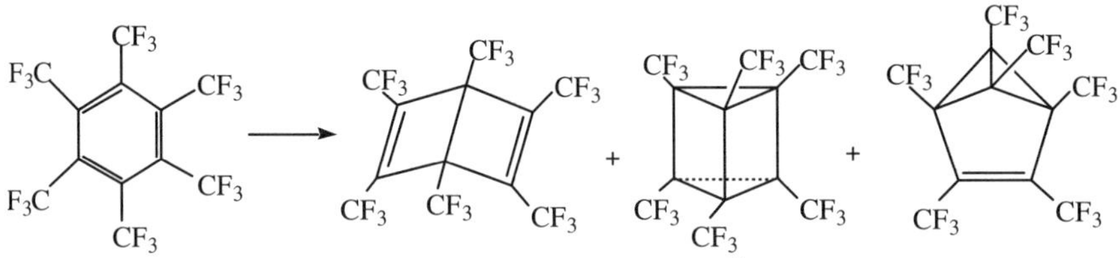

| **F** | **G** | **H** |
| Dewar benzene | prismane | benzvalene |

⟹ EXPLANATION 104

At the first glance, treatment with antimony pentafluoride should be expected to cause addition of fluorine to aromatic rings, or to replace other halogens with fluorine. Under very energetic conditions, bromine is eliminated as a bromide anion and the sextet on carbon 2 causes rearrangement of the napthalene skeleton to an indene skeleton. Addition of two molecules of fluorine from antimony pentafluoride affords perfluoro-3-methylindane, **I** [*141*].

⟶ EXPLANATION 105

At a very high temperature, an intramolecular rearrangement occurs. It starts as the *ortho*-Claisen rearrangement resulting in the formation of a non-aromatic intermediate, 1-(pentafluoro-2-oxocyclohexa-3,5-dienyl)allene. By intramolecular bond shifts, a furan ring is formed to give 2-fluoromethyl-4,5,6,7-tetrafluorobenzofuran [*142*].

References

The numbers in parentheses indicate the location of each reference's citation.

[1] Pauling, L. *The Nature of the Chemical Bond*, 2nd ed.; Cornell University Press: Ithaca, NY, 1948; pp 49, 53. (p. 42).

[2] Glockler, G. *Fluorine Chemistry;* Simons, J. H., ed.; Academic Press: New York, 1950; pp 347–354. (p. 42).

[3] Pauling, L.; Pauling, P. *Chemistry*; Freeman: San Francisco, 1975; pp 740, 741. (p. 42).

[4] Merritt, R. F. *J. Org. Chem.* **1966**, *31*, 3871. (p. 42).

[5] Bornstein, J.; Borden, M. R.; Nunes, F.; Tarlin, H. I. *J. Am. Chem. Soc.* **1963**, *85,* 1609. (p. 42, 46).

[6] Carpenter, W. *J. Org. Chem.* **1966**, *31*, 2688. (p. 43).

[7] Cohen, S.; Kaluszyner, A.; Mechoulam, R. *J. Am. Chem. Soc.* **1957**, *79*, 5979. (p. 43, 44).

[8] Castillon, S.; Dessinges, A.; Faghih, R.; Lukacs, G.; Olesker, A.; Thang, T. T. *J. Org. Chem.* **1985**, *50*, 4913. (p. 45).

[9] Asato, A. E.; Liu, R. S. H. *Tetrahedron Lett.* **1986**, *27*, 3337. (p. 45, 46).

[10] Lacher, J. R.; Kianpour, A.; Park, J. D. *J. Phys. Chem.* **1956**, *60*, 1454. (p. 46).

[11] Hudlický, M. *J. Fluorine Chem.* **1979**, *14,* 189; **1983**, *23*, 241; **1989**, *44*, 345. (p. 46, 47).

[12] Knunyants, I. L.; Krasuskaya, M. P.; Mysov, E. I. *Izv. Akad. Nauk SSSR* **1960,** 1412; *Chem. Abstr.* **1961,** *55*, 349c. (p. 46).

[13] Swarts, F. *Bull. Acad. R. Belg.* **1920**, 399; *Chem. Abstr.* **1922,** *16*, 2316. (p. 46, 47).

[14] Chambers, R. D.; Musgrave, W. K. R.; Drakesmith, F. G. *Br. Pat.* 1134651 (1968); *Chem. Abstr.* **1969**, *70*, 57661. (p. 47).

[15] Banks, R. E.; Haszeldine, R. N.; Latham, J. V.; Young, I. M. *Chem. & Ind. (Lond.)* **1964**, 835; *J. Chem. Soc.* **1965**, 594. (p. 47).

[16] Reinholdt, K.; Margaretha, P. *Helv. Chim. Acta* **1983**, *66*, 2534. (p. 48).

[17] Paleta, O.; Ježek, R.; Dědek, V. *Collect. Czech. Chem. Commun.* **1983**, *48*, 766. (p. 48).

[18] Müller, W.; Walaschewsky, E. *Ger. Pat.* 947364 (1956); *Chem. Abstr.* **1959**, *53*, 4299b. (p. 49).

[19] Hurka, V. R. *U.S. Pat.* 2676983 (1954); *Chem. Abstr.* **1955**, *49*, 5510g. (p. 49).

[20] Haszeldine, R. N.; Nyman, F. *J. Chem. Soc.* **1959**, 1084. (p. 49).

[21] Kobrina, L. S.; Akulenko, N. V.; Yakobson, G. G. *Zh. Org. Khim.* **1972**, *8*, 2375; *Chem. Abstr.* **1973**, *78*, 58326. (p. 50).

[22] Hudlický, M.; Bell, H. M. *J. Fluorine Chem.* **1974**, *4*, 149. (p. 51).

[23] Chambers, R. D.; Musgrave, W. K. R.; Savory, J. *J. Chem. Soc.* **1961**, 3779. (p. 51).

[24] Haszeldine, R. N.; Steele, B. R. *J. Chem. Soc.* **1954**, 3747. (p. 52, 53).

[25] Hudlický, M.; Zahálka, J. *Czech. Pat.* 116341 (1965); *Chem. Abstr.* **1966**, *65*, 2125a; *Br. Pat.* 1070137 (1967). (p. 52).

[26] Hudlický, M. *Czech. Pat.* 116813 (1965); *Chem. Abstr.* **1966**, *65*, 13540h. (p. 52).

[27] Park, J. D.; Sharrah, M. L.; Lacher, J. R. *J. Am. Chem. Soc.* **1949**, *71,* 2339. (p. 52).

[28] Henne, A. L.; Kaye, S. *J. Am. Chem. Soc.* **1950**, *72*, 3369. (p. 53).

[29] Knunyants. I. L.; Shokina, V. V.; Kuleshova, N. D. *Izv. Akad. Nauk SSSR* **1960**, 1693; *Chem. Abstr.* **1961**, *55*, 9254g. (p. 54).

[30] Stacey, F. W.; Harris, J. F., Jr. *J. Org. Chem.* **1962**, *27*, 4089. (p. 54).

[31] Miller, W. T., Jr.; Fried, J. H.; Goldwhite, H. *J. Am. Chem. Soc.* **1960**, *82,* 3091. (p. 54, 55).

[32] Miller, W. T., Jr.; Friedman, M. B.; Fried, J. H.; Koch, H. F. *J. Am. Chem. Soc.* **1961**, *83*, 4105. (p. 55).

[33] Miller, W. T., Jr.; Burnard, R. J. *J. Am. Chem. Soc.* **1968**, *90*, 7367. (p. 55).

[34] Varma, P. S.; Venkat Raman, K. S.; Nilkantiah, P. M. *J. Indian Chem. Soc.* **1944**, *21*, 112. (p. 55).

[35] Illuminati, G.; Marino, G. *J. Am. Chem. Soc.* **1956,** *78,* 4975. (p. 56).

[36] Kumai, S.; Wada, A.; Morikawa, S. *Eur. Pat.* EP355,719 (1990); *Chem. Abstr.* **1990**, *113*, 97175. (p. 56).

[37] Nielsen, A. T.; Chafin, A. P.; Christian, S. L. *J. Org. Chem.* **1984,** *49*, 4575. (p. 56).

[38] Tari, I.; DesMarteau, D. D. *J. Org. Chem.* **1980**, *45*, 1214. (p. 57).

[39] Denivelle, L.; Huynh-Anh-Hoi *Bull. Soc. Chim. Fr.* **1974**, 2171. (p. 57).

[40] Solomon, W. C.; Dee, L. A.; Schults, D. W. *J. Org. Chem.* **1966**, *31*, 1551. (p. 58).

[41] Tiers, G. V. D. *J. Am. Chem. Soc.* **1955**, *77*, 4837. (p. 58).

[42] Tiers, G. V. D. *J. Am. Chem. Soc.* **1955**, *77*, 6704. (p. 58).

[43] Suschitzky, H. *J. Chem. Soc.* **1955**, 4026. (p. 60).

[44] Krespan, C. G. *J. Org. Chem.* **1979**, *44*, 4924. (p. 60).

[45] Galakhov, M. V.; Cherstkov, V. F.; Sterlin, S. R.; German, L. S. *Izv. Akad. Nauk SSSR* **1987,** 958 (Engl. Transl. 886). (p. 61).

[46] Belaventsev, M. A.; Mikheev, L. L.; Pavlov, V. M.; Sokol'skii, G. A.; Knunyants, I. L. *Izv. Akad. Nauk SSSR* **1972,** 2510 (Engl. Transl. 2441). (p. 61).

[47] Knunyants, I. L.; Shokina, V. V.; Mysov, E. I. *Izv. Akad. Nauk. SSSR* **1973,** 2725 (Eng. Transl. 2659). (p. 62).

[48] Cherstkov, V. F.; Sterlin, S. R.; German. L. S.; Knunyants, I. L. *Izv. Akad. Nauk SSSR* **1982**, 2791 (Engl. Transl. 2468). (p. 62)

[49] German, L. S.; Knunyants, I. L.; Sterlin, S. R.; Cherstkov, V. F. *Izv. Akad. Nauk SSSR* **1981**, 1933; *Chem. Abstr.* **1981**, *95*, 203475. (p. 62).

[50] Olah, G.; Kuhn, S. J. *J. Org. Chem.* **1964**, *29*, 2317. (p. 63).

[51] Henne, A. L.; Kraus, D. W. *PhD Thesis*, Ohio State University, 1953. (p. 63).

[52] Kobayashi, Y.; Nagai, T.; Kumadaki, I.; Takahashi, M.; Yamauchi, T. *Chem. Pharm. Bull.* **1984,** *32*, 4382. (p. 64).

[53] Fialkov, Y. A.; Sevast'yan, A, P.; Yagupol'skii, L. M. *Zh. Org. Khim.* **1979**, *15,* 1256 (Engl. Transl. 1121). (p. 64).

[54] Swain, C. G.; Spalding, R. E. T. *J. Am. Chem. Soc.* **1960**, *82,* 6104. (p. 65).

[55] Hudlický, M.; Kraus, E.; Körbl, J.; Čech, M. *Collect. Czech. Chem. Commun.* **1969,** *34,* 833. (p. 65).

[56] Bekker, R. A.; Popkova, V. I.; Knunyants, I. L. *Dokl. Akad. Nauk SSSR* **1978,** *239,* 330 (Engl. Transl. 108). (p. 66).

[57] Stockel, R. F.; Beacham, M. T.; Megson, F. H. *J. Org. Chem.* **1965**, *30,* 1629. (p. 66).

[58] Chambers, R. D.; Hutchinson, J.; Musgrave, W. K. R. *J. Chem. Soc.* **1964,** 5634. (p. 67).

[59] Yagupol'skii, L. M.; Shein, S. M.; Krasnosel'skaya. M. I.; Solodushenkov, S. N. *Zh. Obshch. Khim.* **1965,** *35,* 1261; *Chem..Abstr.* **1965,** *63,* 11417a. (p. 68).

[60] Husted, D. R.; Ahlbrecht, A. H. *J. Am. Chem. Soc.* **1952,** *74,* 5422. (p. 69).

[61] Sykes, A.; Tatlow, J. C.; Thomas, C. R. *Chem. Ind. (Lond.)* **1955,** 630; *J. Chem. Soc.* **1956,** 835. (p. 69).

[62] Simons, J. H.; Ramler, E. O. *J. Am. Chem. Soc.* **1943,** *65,* 389. (p. 69).

[63] Farrah, B. S.; Gilbert, E. E.; Jones, E. S.; Otto, J. A. *J. Org. Chem.* **1965,** *30,* 1006. (p. 69).

[64] Banús, J.; Emeléus, H. J.; Haszeldine, R. N. *J. Chem. Soc.* **1951,** 60. (p. 70).

[65] Dmowski, W. *J. Fluorine Chem.* **1982,** *20,* 589. (p. 70).

[66] Cuthberson, F.; Musgrave, W. K. R. *J. Appl. Chem.* **1957,** *7,* 99. (p. 71).

[67] Iznaden, M.; Portella, C. *J. Fluorine Chem.* **1989,** *43,* 105. (p. 71).

[68] Silversmith, E. F.; Roberts, J. D. *J. Am. Chem. Soc.* **1958,** *80,* 4083. (p. 72).

[69] Kobayashi, Y.; Taguchi, T.; Morikawa, T.; Takase, T.; Takanashi, H. *Tetrahedron Lett.* **1980,** *21,* 1047. (p. 72).

[70] Kozachuk, D. N.; Serguchev, Y. A.; Fialkov, Y. A.; Yagupol'skii, L. M. *Zh. Obshch. Khim.* **1973,** *9,* 1918 (Engl. Transl. 1936). (p. 73).

[71] Park, J. D.; Wilson, L. H.; Lacher, J. R. *J. Org. Chem.* **1963,** *28,* 1008. (p. 73).

[72] Park, J. D.; Dick, J. R.; Adams, J. H. *J. Org. Chem.* **1965,** *30,* 400. (p. 74).

[73] Hudlický, M. Unpublished results. (p. 75).

[74] Knunyants, I. L.; Kildisheva, O. V.; Petrov, I. P. *Zh. Obshch. Khim.* **1949,** *19,* 95; *Chem. Abstr.* **1949,** *43,* 6163. (p. 75).

[75] Platonov, V. E.; Malyuta, N. G.; Yakobson, G. G. *Izv. Akad. Nauk SSSR* **1972,** 2819; *Chem. Abstr.* **1973,** *78,* 83949a. (p. 76).

[76] Hemer, I.; Moravcová, V.; Dědek, V. *Collect. Czech. Chem. Commun.* **1988,** *53,* 619. (p. 76).

[77] Chambers, R. D.; Kirk, J. R.; Powell, R. L. *J. Chem. Soc., Perkin Trans. 1* **1983,** 1239. (p. 77).

[78] Sauvetre, R.; Normant, J.; Villieras, J. *Tetrahedron* **1975,** *31,* 897. (p. 77).

[79] Yamabe, M.; Kumai, S. *U.S. Pat.* 4151200 (1979); *Chem. Abstr.* **1979,** *91,* 38940. (p. 78).

[80] Wakselman, C.; Tordeux, M. *Chem. Commun.* **1984,** 793; *J. Org. Chem.* **1985,** *50,* 4047. (p. 78).

[81] MacNicol, D. D.; Robertson, C. D. *Nature (Lond.)* **1988,** *332,* 59. (p. 80).

[82] Knunyants, I. L.; Struchkov, Y. T.; Bargamova, M. D.; Espenbetov, A. A. *Izv. Akad. Nauk SSSR* **1985,** 1097 (Engl. Transl. 1001); *Chem. Abstr.* **1986,** *104,* 19362. (p. 80, 81).

[83] Yagupol'skii, L. M.; Lukmanov, V. G.; Boiko, V. N.; Alexeeva, L. A. *Zh. Org. Khim.* **1977,** *13,* 2388; *Chem. Abstr.* **1978,** *88,* 62120. (p. 81).

[84] Rozov, L. A.; Mirzabekyants, N. S.; Zeifman, Y. F.; Cheburkov, Y. A.; Knunyants, I. L. *Izv. Akad. Nauk SSSR* 1974, 1355; *Chem. Abstr.* **1974,** *81,* 119875b. (p. 82).

[85] Dvornikova, K. V.; Platonov, V. E.; Yakobson, G. G. *J. Fluorine Chem.* **1985,** *28,* 99. (p. 83).

[86] Bogdanowicz-Szwed, K.; Kawalek, B.; Lieb, M. *J. Fluorine Chem.* **1987,** *35,* 317. (p. 83).

[87] Miller, W. T., Jr.; Fager, E. W.; Griswald, P. H. *J. Am. Chem. Soc.* **1948,** *70,* 431. (p. 85).

[88] Pruett, R. L.; Barr, J. T.; Rapp, K. E.; Bahner, C. T.; Gibson, J. D.;Lafferty, R. H., Jr. *J. Am. Chem. Soc.* **1950,** *72,* 3646. (p. 85, 86).

[89] LaZerte, J. D.; Koshar, R. J. *J. Am. Chem. Soc.* **1955,** *77,* 910. (p. 87).

[90] Banks, R. E.; McGlinchey, M. J. *J. Chem. Soc. C* **1970,** 2172. (p. 87).

[91] Meisenheimer, J. *Ann. Chem. (Justus Liebigs)* **1902,** *323,* 205. (p. 88, 89).

[92] Yakobson, G. G.; Kobrina, L. S.; Vorozhtsov, N. N., Jr. *Zh. Obshch. Khim.* **1965,** *35,* 137; *Chem. Abstr.* **1965,** *62,* 13072h. (p. 88).

[93] Semmelhack, M. F.; Hall, H. T. *Comprehensive Organometallic Chemistry*; Wilkinson, G.; Stone, F. G. A.; Abel, E. W., Eds.; Pergamon Press: Oxford, 1982; pp 381, 382, ref. 55. (p. 88).

[94] Sanger, F. *Biochem. J.* **1945**, *39*, 507. (p. 89).

[95] Bevan, C. W. L. *J. Chem. Soc.* **1951**, 2340. (p. 89, 90).

[96] Hollemann, A. F.; Beekman, J. W. *Recl. Trav. Chim. Pays-Bas* **1904**, *23*, 253. (p. 89, 90).

[97] Kolonko, K. J.; Deinzer, M. L.; Miller, T. L. *Synthesis* **1981**, 133. (p. 90).

[98] Freer, A. A.; MacNicol, D. D.; Mallinson, P. R.; Robertson, C. D. *Tetrahedron Lett.* **1989**, *30*, 5787. (p. 90).

[99] Sket, B.; Zupan, M. *J. Heterocycl. Chem.* **1978**, *15*, 527. (p. 91).

[100] Roe, A.; Montgomery, J. A.; Yarnall, W. A.; Hoyle, V. A., Jr. *J. Org. Chem.* **1956**, *21*, 28. (p. 91).

[101] Bourne, E. J.; Stacey, M.; Tatlow, J. C.; Worrall, R. *J. Chem. Soc.* **1954**, 2006. (p. 92).

[102] Bourne, E. J.; Stacey, M.; Tatlow, J. C.; Worrall, R. *J. Chem. Soc.* **1958**, 3268. (p. 92).

[103] Grillot, G. F.; Aftergut, S.; Marmor, S.; Corrock, F. *J. Org. Chem.* **1958**, *23*, 386. (p. 93).

[104] Burton, D. J.; Herkes, F. E.; Klabunde, K. J. *J. Am. Chem. Soc.* **1966**, *88*, 5042. (p. 94).

[105] Burton, D. J. *J. Fluorine Chem*, **1983**, *23*, 339. (p. 94).

[106] Trabelsi, H.; Rouvier, E.; Gambon, A. *J. Fluorine Chem.* **1986**, *31*, 351. (p. 95).

[107] Knunyants, I. L.; Kocharyan, S. T.; Rokhlin, E. M. *Izv. Akad. Nauk SSSR* **1966**, 1057; *Chem. Abstr.* **1966**, *65*, 12105b. (p. 95).

[108] Elkik, E.; Le Blanc, M.; Far, H. A. *Compt. Rend.* **1971**, *272*, 1895. (p. 96).

[109] Tarrant, P.; Warner, D. A. *J. Am. Chem. Soc.* **1954**, *76*, 1624. (p. 96).

[110] Park, J. D.; Sullivan, R.; McMurtry, R. J. *Tetrahedron Lett.* **1967**, 173. (p. 97).

[111] Drakesmith, F. G.; Richardson, R. D.; Stewart, O. J.; Tarrant, P. *J. Org. Chem.* **1968**, *33*, 286. (p. 98).

[112] Yagupol'skii, L. M.; Cherednichenko, P. G.; Kremlev, M. M. *Zh. Org. Khim.* **1987**, *23*, 279 (Engl. Transl. 246); *Chem. Abstr.* **1987**, *107*, 236795. (p. 99).

[113] Pošta, A.; Paleta, O. *Collect. Czech. Chem. Commun.* **1972**, *37*, 3946. (p. 99).

[114] Fujita, M.; Hiyama, T. *Tetrahedron Lett.* **1986**, *27*, 3655. (p. 99, 100).

[115] Clark, J. H.; McClinton, M. A.; Jones, C. W.; Landon, P.; Bishop, D.; Blade, R. J. *Tetrahedron Lett.* **1989**, *30*, 2133. (p. 100).

[116] Walborsky, H. M.; Schwarz, M. *J. Am. Chem. Soc.* **1953**, *75*, 3241. (p. 101).

[117] Knunyants, I. L.; Cheburkov, Y. A. *Izv. Akad. Nauk SSSR* **1960**, 1516; *Chem. Abstr.* **1961**, *55*, 1462d. (p. 101).

[118] Bernett, W. A. *J. Org. Chem.* **1969**, *34*, 1772. (p. 102).

[119] Bartlett, P. D.; Montgomery, L. K.; Seidel, B. *J. Am. Chem. Soc.* **1964**, *86*, 616. (p. 102).

[120] Bartlett, P. D. *Quart. Rev.* **1970**, *24*, 473. (p. 102.)

[121] Sargeant, P. B. *J. Org. Chem.* **1970**, *35*, 678. (p. 103).

[122] Banks, R. E.; Barlow, M. G.; Haszeldine, R. N.; Morton, W. D. *J. Chem. Soc., Perkin Trans. 1*, **1974**, 1266. (p. 103).

[123] Bartlett, P. D.; Jacobson, B. M.; Walker, L. E. *J. Am. Chem. Soc.* **1973**, *95*, 146. (p. 104).

[124] Burger, K.; Thenn, W.; Rauh, R.; Schickaneder, H.; Gieren, A. *Chem. Ber.* **1975**, *108*, 1460. (p. 104).

[125] Burger, K.; Schickaneder, H.; Thenn, W. *Tetrahedron Lett.* **1975**, 1125. (p. 104).

[126] Hees, U.; Ledermann, M.; Regitz, M. *Syn. Lett.* **1990,** 401. (p. 104).

[127] Prober, M.; Miller, W. T., Jr. *J. Am. Chem. Soc.* **1949,** *71,* 598. (p. 105).

[128] Karle, I. L.; Karle, J.; Owen, T. B.; Broge, R. W.; Fox, A. H.; Hoard, J. L. *J. Am. Chem. Soc.* **1964,** *86,* 2523. (p. 105).

[129] Kobayashi, Y.; Morikawa, A.; Yoshizawa, A.; Taguchi, T. *Tetrahedron Lett.* **1981,** *22,* 5297. (p. 105).

[130] Hudlický, M. *J. Fluorine Chem.* **1984,** *25,* 353. (p. 106).

[131] Lee, J. G.; Bartsch, R. A. *J. Am. Chem. Soc.* **1979,** *101,* 228. (p. 107).

[132] Hudlický, M. *J. Fluorine Chem.* **1986,** *32,* 441. (p. 107).

[133] Hudlický, M.; Lejhancová, I. *Collect. Czech. Chem. Commun.* **1963,** *28,* 2455. (p. 108).

[134] Miller, W. T., Jr.; Fager, E. W.; Griswald, P. H. *J. Am. Chem. Soc.* **1950,** *72,* 705. (p. 108, 109).

[135] Madai, H. G.; Müller, R. *J. Prakt. Chem.* **1963,** *19,* 83. (p. 109).

[136] Henne, A. L.; Newby, T. H. *J. Am. Chem. Soc.* **1948,** *70,* 130. (p. 109).

[137] Makarov, S. P.; Yakubovich, A. Y.; Dubov, S. S.; Medvedev, A. N. *Dokl. Akad. Nauk SSSR* **1965,** *160,* 1319; *Chem. Abstr.* **1965,** *62,* 14481f. (p. 109).

[138] Barr, D. A.; Haszeldine. R. N. *J. Chem. Soc.* **1957,** 30. (p. 110).

[139] Camaggi, G.; Gozzo, F.; Cevidalli, G. *Chem. Commun.* **1966,** 313. (p. 110).

[140] Lemal, D. M.; Staros, J. V.; Austel, V. *J. Am. Chem. Soc.* **1969,** *91,* 3373. (p. 110).

[141] Pozdnyakovich, Y. V.; Bardin, V. V.; Shtark, A. A.; Shteingarts, V. D. *Zh. Org. Khim.* **1979,** *15,* 656 (Engl. Transl. 583). (p. 111).

[142] Brooke, G. M.; Wallis, D. I. *J. Chem. Soc., Perkin Trans. 1,* **1981,** 1417. (p. 111).

Author Index

The numbers following the author's names refer to the List of References (pp. 113–117), and the number in parentheses refers to the text page on which the reference is cited.

Adams, J. H., 72 (74)
Aftergut, S., 103 (93)
Ahlbrecht, A. H., 60 (69)
Akulenko, N. V., 21 (50)
Alexeeva, L. A., 83 (81)
Asato, A. E., 9 (45, 46)
Austel, V., 140 (110)

Bahner, C. T., 88 (85, 86)
Banks, R. E., 15 (47), 90 (87), 122 (103)
Banús, J., 64 (70)
Bardin, V. V., 141 (111)
Bargamova, M. D., 82 (80, 81)
Barlow, M. G., 122 (103)
Barr, D. A., 138 (110)
Barr, J. T., 88 (85, 86)
Bartlett, P. D., 119 (102), 120 (102), 123 (104)
Bartsch, R. A., 131 (107)
Beacham, M. T., 57 (66)
Beckman, J. W., 96 (89, 90)
Bekker, R. A., 56 (66)
Belaventsev, M. A., 46 (61)
Bernett, W. A., 118 (102)
Bevan, C. W. L., 95 (89, 90)
Bishop, D., 115 (100)
Blade, R. J., 115 (100)
Bogdanowicz-Szwed, K., 86 (83)
Boiko, V. N., 83 (81)
Borden, M. R., 5 (42, 46)
Borne, E. J., 101 (92), 102 (92)
Bornstein, J., 5 (42, 46)
Broge, R. W., 128 (105)
Brooke, G. M., 142 (111)
Burger, K., 124 (104), 125 (104)
Burnard, R. J., 33 (55)
Burton, D. J., 104 (94), 105 (94)

Carnaggi, G., 139 (110)
Carpenter, W., 6 (43)
Castillon, S., 8 (45)
Cevidalli, G., 139 (110)
Chafin, A. P., 37 (56)
Chambers, R. D., 14 (47), 23 (51), 58 (67), 77 (77)
Cheburkov, Y. A., 84 (82), 117 (101)
Cherednichenko, P. G., 112 (99)
Cherstkov, V. F., 45 (61), 48 (62), 49 (62)
Christian, S. L., 37 (56)
Clark, J. H., 115 (100)
Cohen, S., 7 (43, 44)
Corrock, F., 103 (93)
Cuthberson, F., 66 (71)

Dědek, V., 17 (48), 76 (76)
Dee, L. A., 40 (58)
Deinzer, M. L., 97 (90)
Denivelle, L., 39 (57)
DesMarteau, D. D., 38 (57)
Dessinges, A., 8 (45)
Dick, J. R., 72 (74)
Dmowski, W., 65 (70)
Drakesmith, F. G., 14 (47), 111 (98)
Dubov, S. S., 137 (109)
Dvornikova, K. V., 85 (83)

Elkik, E., 108 (96)
Emeléus, H. J., 64 (70)
Espenbetov, A. A., 82 (80, 81)

Fager, E. W., 87 (85), 134 (108,109)
Faghih, R., 8 (45)
Far, H. A., 108 (96)
Farrah, B. S., 63 (69)
Fialkov, Y. A., 53 (64), 70 (73)
Fox, A. H., 128 (105)
Freer, A. A., 98 (90)
Fried, J. H., 31 (54, 55), 32 (55)
Friedman, M. B., 32 (55)
Fujita, M., 114 (99, 100)

Galakhov, M. V., 45 (61)
Gambon, A., 106 (95)
German, L. S., 45 (61), 48 (62), 49 (62)
Gibson, J. D., 88 (85, 86)
Gieren, A., 124 (104)
Gilbert, E. E., 63 (69)
Glockler, G., 2 (42)
Goldwhite, H., 31 (54, 55)
Gozzo, F., 139 (110)
Grillot, G. F., 103 (93)
Griswald, P. H., 87 (85), 134 (108 109),

Hall, H. T., 93 (88)
Harris, J. F., Jr., 30 (54)
Haszeldine, R. N., 15 (47), 20 (49), 24 (52, 53), 64 (70), 122 (103), 138 (110)
Hees, U., 126 (104)
Hemer, I., 76 (76)
Henne, A. L., 28 (53), 51 (63), 136 (109)
Herkes, F. E., 104 (94)
Hiyama, T., 114 (99, 100)
Hoard, J. L., 128 (105)
Hollermann, A. F., 96 (89, 90)
Hoyle, V. A., Jr., 100 (91)

Hudlický, M., 11 (46, 47), 22 (51), 25 (52), 26 (52), 55 (65), 73 (75), 130 (106), 132 (107), 133 (108)
Hurka, V. R., 19 (49)
Husted, D. R., 60 (69)
Hutchison, J., 58 (67)
Huynh-Anh-Hoi, 39 (57)

Illuminati, G., 35 (56)
Iznaden, M., 67 (71)

Jacobson, B. M., 123 (104)
Ježek, R., 17 (48)
Jones, C. W., 115 (100)
Jones, E. S., 63 (69)

Kaluszyner, A., 7 (43, 44)
Karle, I. L., 128 (105)
Karle, J., 128 (105)
Kawalek, B., 86 (83)
Kaye, S., 28 (53)
Kianpour, A., 10 (46)
Kildisheve, O. V., 74 (75)
Kirk, J. R., 77 (77)
Klabunde, K. J., 104 (94)
Knunyants, I. L., 12 (46), 29 (54), 46 (61), 47 (62), 48 (62), 49 (62), 56 (66), 74 (75), 82 (80, 81), 84 (82), 107 (95), 117 (101)
Kobayashi, Y., 52 (64), 69 (72), 129 (105)
Kobrina, L. S., 21 (50), 92 (88)
Koch, H. F., 32 (55)
Kocharyan, S. T., 107 (95)
Kolonko, K. J., 97 (90)
Körbil, J., 55 (65)
Koshar, R. J., 89 (87)
Kozachuk, D. N., 70 (73)
Krasnosel'skaya, M. L., 59 (69)
Krasuskaya, M. P., 12 (46)
Kraus, D. W., 51 (63)
Kraus, E., 55 (65)
Kremlev, M. M., 112 (99)
Krespan, C. G., 44 (60)
Kuhn, S. J., 50 (63)
Kuleshova, N. D., 29 (54)
Kumadaki, I., 52 (64)
Kumai, S., 36 (56), 79 (78)

Lacher, J. R., 10 (46), 27 (52), 71 (73)
Lafferty, R. H., Jr., 88 (85, 86)
Landon, P., 115 (100)
Latham, J. V., 15 (47)
LaZerte, J. D., 89 (87)
Le Blanc, M., 108 (96)
Ledermann, M., 126 (104)
Lee, J. G., 131 (107)
Lejhancová, I., 133 (108)
Lemal, D. M., 140 (110)
Lieb, M., 86 (83)
Liu, R. S. H., 9 (45, 46)
Luckacs, G., 8 (45)
Lukmanov, V. G., 83 (81)

MacNicol, D. D., 81 (80), 98 (90)
Madai, H. G., 135 (109)
Makarov, S. P., 137 (109)
Mallinson, P. R., 98 (90)
Malyuta, N. G., 75 (76)

Margaretha, P., 16 (48)
Marino, G., 35 (56)
Marmor, S., 103 (93)
McClinton, M. A., 115 (100)
McGlinchey, M. J., 90 (87)
McMurtry, R., 110 (97)
Mechoulam, R., 7 (43, 44)
Medvedev, A. N., 137 (109)
Megson, F. H., 57 (66)
Meisenheimer, J., 91 (88, 89)
Merritt, R.F., 4 (42)
Mikheev, L. L., 46 (61)
Miller, T. L., 97 (90)
Miller, W. T., Jr., 31 (54, 55), 32 (55), 33 (55), 87 (85), 109), 127 (105), 134 (108
Mirzabekyants, N. S., 84 (82)
Montgomery, J. A., 100 (91)
Montgomery, L. K., 119 (102)
Moravcova, V., 76 (76)
Morikawa, A., 129 (105)
Morikawa, S., 36 (56)
Morikawa, T., 69 (72)
Morton, W. D., 122 (103)
Müller, R., 135 (109)
Müller, W., 18 (49)
Musgrave, W. K. R., 14 (47), 23 (51), 58 (67), 66 (71)
Mysov, E. I., 12 (46), 47 (62)

Nagai, T., 52 (64)
Newby, T. H., 136 (109)
Nielsen, A. T., 37 (56)
Nilkantiah, P. M., 34 (55)
Normant, J., 78 (77)
Nunes, F., 5 (42, 46)
Nyman, F., 20 (49)

Olah, G., 50 (63)
Olesker, A., 8 (45)
Otto, J. A., 63 (69)
Owen, T. B., 128 (105)

Paleta, O., 17 (48), 113 (99)
Park, J. D., 10 (46), 27 (52), 71 (73), 72 (74), 110 (97)
Pauling, L., 1 (42), 3 (42)
Pauling, P., 3 (42)
Pavlov, V. M., 46 (61)
Petrov, I. P., 74 (75)
Platonov, V. E., 75 (76), 85 (83)
Popkova, V. I., 56 (66)
Portella, C., 67 (71)
Pošta, A., 113 (99)
Powell, R. L., 77 (77)
Pozdnyakovich, Y. V., 141 (111)
Prober, M., 127 (105)
Pruett, R. L., 88 (85, 86)

Ramler, E. O., 62 (69)
Rapp, K. E., 88 (85, 86)
Rauh, R., 124 (104)
Regitz, M., 126 (104)
Reinholdt, K., 16 (48)
Richardson, R. D., 111 (98)
Roberts, J. D., 68 (72)
Robertson, C. D., 81 (80), 98 (90)
Roe, A., 100 (91)

Rokhlin, E. M., 107 (95)
Rouvier, E., 106 (95)
Rozov, L. A., 84 (82)

Sanger, F., 94 (89)
Sargeant, P. B., 121 (103)
Sauvetre, R., 78 (77)
Savory, J., 23 (51)
Schickaneder, H., 124 (104), 125 (104)
Schults, D. W., 40 (58)
Schwarz, M., 116 (101)
Seidel, B., 119 (102)
Semmelhack, M. F., 93 (88)
Serguchev, Y. A., 70 (73)
Sevast'yan, A. P., 53 (64)
Sharrah, M. L., 27 (52)
Shein, S. M., 59 (69)
Shokina, V. V., 29 (54), 47 (62)
Shtark, A. A., 141 (111)
Shteingarts, V. D., 141 (111)
Silversmith, E. F., 68 (72)
Simons, J. H., 62 (69)
Sket, B., 99 (91)
Sokol'skii, G. A., 46 (61)
Solodushenkov, S. N., 59 (69)
Solomon, W. C., 40 (58)
Spalding, R. E. T., 54 (65)
Stacey, F. W., 30 (54)
Stacey, M., 101 (92), 102 (92)
Staros, J. V., 140 (110)
Steele, B. R., 24 (52, 53)
Sterlin, S. R., 45 (61), 48 (62), 49 (62)
Stewart, O. J., 111 (98)
Stockel, R. F., 57 (66)
Struchkov, Y. T., 82 (80, 81)
Sullivan, R., 110 (97)
Suschitzky, H., 43 (60)
Swain, C. G., 54 (65)
Swarts, F., 13 (46, 47)
Sykes, A., 61 (69)

Taguchi, T., 69 (72), 129 (105)
Takahashi, M., 52 (64)

Takanashi, H., 69 (72)
Takase, T., 69 (72)
Tari, I., 38 (57)
Tarlin, H. I., 5 (42, 46)
Tarrant, P., 109 (96), 111 (98)
Tatlow, J. C., 61 (69), 101 (92), 102 (92)
Thang, T. T., 8 (45)
Thenn, W., 124 (104), 125 (104)
Thomas, C. R., 61 (69)
Tiers, G. V. D., 41 (58), 42 (58)
Tordeux, M., 80 (78)
Trabelsi, H., 106 (95)

Varma, P. S., 34 (55)
Venkat Raman, K. S., 34 (55)
Villieras, J., 78 (77)
Vorozhtsov, N. N., Jr., 92 (88)

Wada, A., 36 (56)
Wakselman, C., 80 (78)
Walaschewsky, E., 18 (49)
Walborsky, H. M., 116 (101)
Walker, L. E., 123 (104)
Wallis, D. I., 142 (111)
Warner, D. A., 109 (96)
Wilson, L. H., 71 (73)
Worrall, R., 101 (92), 102 (92)

Yagupol'skii, L. M., 53 (64), 59 (69), 70 (73), 83
    (81), 112 (99)
Yakobson, G. G., 21 (50), 75 (76), 85 (83), 92
    (88)
Yakubovich, A. Y., 137 (109)
Yamabe, M., 79 (78)
Yamauchi, T., 52 (64)
Yarnall, W. A., 100 (91)
Yoshizawa, A., 129 (105)
Young, I. M., 15 (47)

Zahalka, J., 25 (52)
Zeifman, Y. F., 84 (82)
Zupan, M., 99 (91)

Subject Index

Acetylacetone, reaction with diethylaminosulfur trifluoride, 46
1-Acetyl-2,2-difluoro-3-phenylcyclopropane, hydrolysis, 18, 72
Acetylene, reaction with hexafluoroacetone azine, 32, 104
Acid-catalyzed additions and substitutions, 14, 15, 62–64
Acidic media, effect on rate of hydrolysis of benzyl fluoride, 65
Acrolein, Michael addition, 95
Acrylonitrile, reaction with tris(trifluoromethyl)-methane, 28
Acylation, with acyl trifluoroacetates, 92, 93
Acylations, 26, 92, 93
Acyl fluoride, by hydrolysis of trifluoromethyl group, 68, 69
Acylium cation, from acyl trifluoroacetate, 92, 93
Acyl trifluoroacetates, acylating agents, 26, 92, 93
Addition of fluorine to diphenylethylenes, 4
Addition of hydrogen bromide to chlorotrifluoro-ethylene, 8
Additions, 31–33, 101–105
  electrophilic, 54
  free-radical, 86, 87, 102–104
  nucleophilic to fluoroalkenes, 22–24
  of sulfur trioxide, 60, 61
  to trifluoromethyl compounds, 101, 102
Aldehydes, reaction with 1,1,1-trichlorotrifluoro-ethane and zinc, 30
Aldol-type condensations, 27, 28
Alkali thiophenoxides, reaction with polyfluoro-halomethanes, 21
Alkenes, fluorinated, reaction with alcohols, 24
octakis(Alkoxy)naphthalene, from perfluoro-naphthalene, 90, 91
Alkylations, 19–24, 73–87
Alkyl fluoroalkyl ethers, 84
Allylic fluorines
  in perfluoro-3,4-dimethyl-3-hexene, 76
  reactivity, 72
Aluminum bromide
  catalyst for addition of hydrogen bromide, 53
  catalyst for rearrangements, 108, 109
  replacement of fluorine by bromine, 58
Aluminum chloride
  catalyst for addition of hydrogen halides, 53
  catalyst for rearrangements, 108, 109
  catalyst in Friedel-Crafts synthesis, 63
  reaction with fluorohaloethanes, 35
  reaction with perfluoro-2-butyltetrahydro-furan, 11
  reaction with perfluorotetrahydropyran, 12
  reaction with 1-phenylperfluoropropene, 14
  replacement of fluorine by chlorine, 58, 64

Aluminum halides, reaction with perfluorocyclo-alkenes, 11
Amines, reaction with chlorofluoroalkenes, 23
2-Aminoethanol
  reaction with ethyl bis(trifluoro)methyl acrylate, 31, 101,102
  reaction with ethyl 4,4,4-trifluoro-3-trifluoro-methyl-2-butenoate, 31, 101, 102
Amino group, activation of trifluoromethyl group for hydrolysis, 68
2-Amino-4-trifluoromethylbenzoic acid, from 2,5-bis(trifluoromethyl)aniline, 68
Ammonia
  reaction with trifluoromethyl group, 22
  reaction with trifluoronitrosomethane, 35, 109
Aniline, reaction with chlorotrifluoroethylene, 23, 85
"Anti-Markovnikov" addition, 53, 54
Antimony dichlorotrifluoride, reaction with 3-chloroperfluoropropene, 109
Antimony halides, reaction with 3-chloroper-fluoropropene, 35
Antimony pentafluoride, reaction with 2-bromo-perfluoronaphthalene, 36, 110, 111
Arylations, 24–26, 88–91
octakis(Aryloxy)naphthalene, from perfluoro-naphthalene, 90, 91
Azide anion, reaction with 1H-pentafluoropro-pene, 87
1-Azido-1,2,3,3,3-pentafluoropropane, from 1H-pentafluoropropene, 87
1-Azido-2,3,3,3-tetrafluoropropane, from 1H-pentafluoropropene and azide anion, 87

Back-donation of electrons
  by fluorine, 47, 51, 52, 55, 60, 61, 70, 84, 85
  by fluorine to electrophilic carbon, 51
  by oxygen, 73
  from phenolic oxygen, 70
Benzaldehyde, reaction with dichlorotrifluoro-ethylzinc chloride, 98–100
Benzene
  reaction with chloro-2-fluoropropanes, 14
  reaction with 2-chloro-1,1,1-trifluoropropane, 63
  reaction with trifluoropropene, 14
Benzophenone, sensitizer for ultraviolet irradia-tion, 91
Benzoyl peroxide, initiator of free-radical reac-tions, 86
Benzoyl trifluoroacetate, benzoylating agent, 92, 93
Benzyl chloride, hydrolysis, 65
Benzyl fluoride, hydrolysis, 65
Benzyl halides, hydrolysis, 15
Benzylic fluorine, elimination, 70

Bond dissociation energy
  carbon-chlorine bond, 63
  fluorine, 42
  halogens, 3
  hydrogen fluoride, 42
Boron trifluoride, catalyst in Friedel-Crafts
  synthesis, 62
Bromine, atomic, attacking species in additions,
  52, 53
Bromine fluoride, from bromine trifluoride and
  bromine, 51
1-Bromo-2-chloro-1,1,2-trifluoroethane, reaction
  with aluminum chloride, 35
2-Bromo-4,5-difluoronitrobenzene, reaction with
  sodium sulfide, 26, 91
4-Bromo-3-ethoxyhexafluorobutene, from 1,4-
  dibromohexafluoro-2-butene, 76
2-Bromo-1-ethoxytetrafluorocyclobutene, from
  2-bromo-1-chlorotetrafluorocyclobutene,
  74
1-Bromo-2-fluorocyclohexane, dehydrohalogenation, 107, 108
1-Bromo-2-fluorocyclohexanes, reaction with
  bases, 34
bis(3-Bromo-6-fluoro-4-nitrophenyl)sulfide, from
  2-bromo-4,5-difluoronitrobenzene, 91
Bromofluorosuccinates, reaction with potassium
  acetate, 34, 106
4-Bromopentafluorocyclohexa-2,5-dienone, from
  pentafluorophenol and tert-hypobromite,
  57
N-Bromoperfluorobutyramide, from perfluorobutyramide, 110
2-Bromoperfluoronaphthalene
  reaction with antimony pentafluoride, 36
  rearrangement, 111
1-Bromo-2-tetrafluorocyclobutene, reaction with
  potassium hydroxide and ethanol, 19
Bromotrifluoroethylene, reaction with butyllithium, 30
1,3-Butadiene
  reaction with fluoroalkenes, 31, 32, 102–104
  reaction with perfluorovinylsulfur pentafluoride, 32, 103
1,3-Butadienes, perfluoro, dimerization, 33, 105
tert-Butyl alcohol, solvent in hydrolysis, 66
tris(tert-Butyl)azete, reaction with trifluoroacetonitrile, 33, 104
tert-Butyl hypobromite, reaction with pentafluorophenol, 11, 57
Butyllithium
  reaction with chlorofluoroalkenes, 29, 30, 98,
  99
  reaction with fluoroalkenes, 29–30, 98, 99
  reaction with trifluoroethylene, 30, 98

Carbocations, rearrangements, 63, 64
Carbon-carbon bond cleavage, 63, 111
  by lead tetraacetate, 49, 50
Carbon-chlorine bond, dissociation energy, 63
Carbon-fluorine bond, hydrogenolysis, 46–48
Carbon-iodine bond, polarization in perfluoro-
  alkyl iodides, 69, 70
Carbon-nitrogen bond, cleavage by chlorine, 56
Carbonyl compounds, reaction with phosphines, 94
Catalyst, boron trifluoride in Friedel-Crafts syn-
  thesis, 62, 63

Catalytic hydrogenation
  of alkyl fluorides, 46, 47
  of chlorotrifluoropyridine, 48
  of fluoro compounds, 6, 46, 47
  of fluorofumaric and fluoromaleic acids, 6, 46,
  47
  of trans-α-fluorostilbene, 6
  simultaneous (conjugate), 47
Cesium fluoride
  in nucleophilic addition of hydrogen fluoride,
  54, 55
  in opening perfluoro-γ-valerolactone, 78
Chichibabin reaction, 67
Chloride anion, as a nucleophile, 56
Chlorination
  of 3,4-difluoronitrobenzene, 10
  of o-fluorotoluene, 9, 55
Chlorine
  atomic form, 56
  cleavage of carbon-nitrogen bond, 56
Chlorine cation, addition to fluoroalkenes, 57
1-Chloro-1,1-dibromotrifluoroethane, from
  1-chloro-1,2-dibromotrifluoroethane, 109
1-Chloro-2,4-diethoxy-3,3-difluorocyclobutene,
  from 1-chloro-3,3,4,4-tetrafluorocyclo-
  butene, 74
2-Chloro-3,3-diethylpentafluorocyclopentene,
  from 1,2-dichlorohexafluorocyclopen-
  tene, 97
Chlorodifluoroacetyl fluoride
  from chlorotrifluoroethylene, 49
  from chlorotrifluoroethylene oxide, 49
1-Chloro-3,4-difluorobenzene, from 3,4-difluoro-
  1-nitrobenzene, 56
1-Chloro-1,2-difluoro-2-ethoxyethylene, from
  chlorotrifluoroethylene, 77
1-Chloro-1,2-difluoroethylene, reaction with
  butyllithium, 30, 98
Chlorodifluoromethane
  formation of difluorocarbene, 83
  reaction with pentafluorophenol, 20
  reaction with perfluorobenzotrichloride, 22, 83
Chlorodifluoromethyl phenyl sulfide, from
  thiophenol and difluorocarbene, 78
3-Chloro-2,3-difluoro-1-phenyl-1-azapropene,
  from chlorotrifluoroethylene, 85
2-Chloro-3,3-difluoro-1-phenylpropen-2-ol, from
  dichlorotrifluoroethylzinc chloride, 100
2-Chloro-1,1-difluoro-1-trifluoroacetoxyethane,
  from 1,1-difluoroethylene, 57
2-Chloro-1-ethoxytetrafluorocyclobutene, from
  1-bromo-1-chlorotetrafluorocyclobutene,
  74
1-Chloro-2-ethoxy-2,3,3-trifluorocyclobutene,
  from 1-chloro-3,3,4,4-tetrafluorocyclo-
  butene, 74
1-Chloro-2-ethylhexafluorocyclopentene, from
  1,2-dichlorohexafluorocyclopentene, 97
Chlorofluoroacetic acid, N,N'-diphenylamidine,
  85
Chlorofluoroalkenes and cycloalkenes
  reaction with alcohols, 19, 20, 23, 73, 74, 76,
  77
  reaction with amines, 23, 85, 86
  reaction with hydrazine, 22, 80, 81
1-Chloro-2-fluoroethane, reaction with ethyl
  acetoacetate, 19, 75

Chlorofluoromethyl phenyl ketone
  by Grignard synthesis, 95, 96
  reaction with methylmagnesium bromide, 95,
    96
1-Chloro-2-fluoropropane
  by Friedel-Crafts synthesis, 63
  reaction of benzene with, 14, 62, 63
Chlorofluoropyridines
  hydrolysis, 16, 67, 68
  reduction, 6, 47, 48
2-Chloro-6-fluorotoluene, by chlorination of
    o-fluorotoluene, 55
5-Chloro-2-fluorotoluene, by chlorination of
    o-fluorotoluene, 55
3-Chloropentafluoropropene, reaction with
    phenylmagnesium bromide, 29, 96
Chloropentanitrobenzene, from fluoropentanitro-
    benzene, 56
3-Chloroperfluoropropene
  reaction with antimony halides, 35, 109
  reaction with phenylmagnesium bromide, 29,
    96
  rearrangement to 1-chloroperfluoropropene,
    109
1-Chloroperfluoropropene, from 3-chloroper-
    fluoropropene, 109
α-Chloroperfluorostyrene, from perfluorobenzo-
    trichloride, 83
2-Chloro-2-phenylpropanal, from ethyl chloro-
    fluoroacetate, 28, 96
3-Chloro-1-phenyltetrafluoropropene, from
    3-chloropentafluoropropene, 96
bis(*p*-Chlorophenyl)-2,2,2-trichloroethane
    (DDT), 43
2-Chloro-1-phenyl-3,3,3-trifluoropropene, from
    dichlorotrifluoroethylzinc chloride, 99
1-Chloro-3,3,4,4-tetrafluorocyclobutene
  reaction with ethanol, 74
  reaction with potassium hydroxide, 19
3-Chlorotetrafluoropyridine
  hydrogenation, 6, 48
  reduction, 6, 47, 48
  reduction with hydrides, 6, 48
Chlorotrifluoroethylene
  oxidation with oxygen, 7, 49
  polarization of the double bond, 52, 77
  reaction with aniline, 23, 85
  reaction with dimethylamine, 23, 85, 86
  reaction with ethanol, 77
  reaction with hydrogen bromide, 8, 53
  reaction with iodine fluoride, 51, 52
  reaction with iodine pentafluoride and iodine, 8
  reaction with sodium ethoxide, 20, 77
Chlorotrifluoroethylene oxide, by oxidation of
    chlorotrifluoroethylene, 49
2-Chloro-1,1,1-trifluoropropane, reaction with
    benzene, 14, 63
3-Chloro-2,5,6-trifluoropyridine, from 3-chloro-
    tetrafluoropyridine, 47, 48
Chromium tricarbonyl fluorobenzene, conver-
    sion to chromium tricarbonyl cyanoben-
    zene, 88
Claisen rearrangement, of fluoroaromatic acetyl-
    enic ethers, 37, 111
Complex hydrides, reduction with, 47
Conjugate hydrogenation, of unsaturated fluoro
    compounds, 47

Conjugated dienes, additions, 102–104
Copper, reaction with dibromodifluoromethane,
    31
"Crisscross" cycloaddition, of hexafluoroacetone
    azine, 104
18-Crown-6-ether, catalyst for nucleophilic
    displacement, 25
Cycloaddition
  (2+2), 31, 32, 102–104
  (2+3), 32, 104
  (2+4), 32, 104
  (4+2), 104, 105
  (2+2) retro, 104, 105
  (2+2) to form four-membered rings, 102
Cycloaddition mechanism of fluoroalkenes
  biradical, 102
  concerted, 102
Cyclobutanetetronetetrakis(hydrazone), from per-
    fluorocyclobutene and hydrazine, 80
Cyclobutene, perfluoro- or hexafluoro-, 22
Cyclohexanone, by reduction of 2-fluorocyclo-
    hexane, 48
Cyclopentanepentonepentakis(phenylhydra-
    zone), from 1,2-dichlorohexafluorocyclo-
    pentene, 81

DAST. *See* Diethylaminosulfur trifluoride
    (DAST)
DDT. *See* 1,1-bis(p-Chlorophenyl)-2,2,2-
    trichloroethane (DDT)
Dehydrogenation, of –CH–NH group, 81, 82
Dehydrohalogenation, of bromofluorocyclohex-
    anes, 107, 108
1,3-Diamino-4-nitro-2,5,6-trifluorobenzene, oxi-
    dation to 1,3-diazo-5-fluoro-4-nitro-2,6-
    benzenedioxide, 8, 50, 51
1,2-Diaminotetrafluorobenzene, cleavage to
    dinitrile of tetrafluoro-2,4-hexanedioic
    acid, 7, 50
1,3-Diazo-5-fluoro-4-nitro-2,6-dioxide, 51
1,3-Diazonium-4-nitro-2,5,6-trifluorobenzene,
    oxidation to 1,3-diazo-5-fluoro-4-nitro-
    2,6-benzenedioxide, 51
Diazotization, of fluoroaromatic diamine, 50
Dibenzoyl peroxide, initiator, 91
Dibromodifluoromethane
  reaction with 4-chloro-3-nitrobenzotrifluoride
    and copper, 31, 100
  reaction with tris(dimethylamino)phosphine
    and fluorinated ketones, 27, 94
Dibromotris(dimethylamino)phosphorane, from
    tris(dimethylamino)phosphine, 94
1,4-Dibromohexafluoro-2-butene, reaction with
    sodium ethoxide, 20, 76
Dibutylhydroxyphosphine, from tributylphos-
    phine, 94
Dibutyloxophosphorane, from tributylphosphine,
    94
β-Dicarbonyl compounds, reaction with diethyl-
    aminosulfur trifluoride, 5, 45, 46
1,1-Dichlorodifluoroethylene
  reaction with 1,3-butadiene, 31, 102
  reaction with methanol, 23, 84
1,2-Dichlorodifluoroethylene
  reaction with difluorocarbene, 103
  reaction with perfluoropropylene oxide, 32,
    103

1,1-Dichloro-2,2-difluoroethylene, reaction with
   methanol, 84
1,1-Dichloro-2,2-difluoroethyl methyl ether, from
   1,1-dichloro-2,2-difluoroethylene, 84
2,2-Dichloro-1,1-difluoroethyl methyl ether, from
   1,1-dichloro-2,2-difluoroethylene, 84
3,5-Dichloro-2,6-difluoro-4-hydroxypyridine,
   from 3,5-dichloro-2,4,6-trifluoropyri-
   dine, 68
3,5-Dichloro-4,6-difluoro-1-hydroxypyridine,
   from 3,5-dichloro-2,4,6-trifluoropyri-
   dine, 68
Dichlorodifluoromethane, reaction with alkali
   thiophenoxides, 21, 78
1,2-Dichloro-1-fluoroethylene, reaction with
   butyllithium and acetone, 29, 30, 98
3,4-Dichloro-4-fluoro-2-methyl-3-buten-2-ol,
   from 1,2-dichlorofluoroethylene, 98
1,2-Dichloro-2-fluorovinyldimethylcarbinol,
   from 1,2-dichlorofluoroethylene, 98
2,2-Dichloro-1-fluorovinyl methyl ether, from
   1,1-dichloro-2,2-difluoroethylene, 23, 84
1,2-Dichlorohexabromocyclopentene, from
   1,2-dichlorohexafluorocyclopentene, 58
1,2-Dichlorohexafluorocyclopentene
   reaction with aluminum bromide, 58
   reaction with ethylmagnesium bromide, 29, 97
1,2-Dichloroperfluorocyclopropane, from
   1,2-dichlorodifluoroethylene, 103
3,5-Dichlorotrifluoropyridine, hydrolysis, 16, 67
1,1-Dichloro-2,2,2-trifluoroethylzinc chloride
   formation from 1,1,1-trichlorotrifluoroethane,
   98–100
   reaction with benzaldehyde, 98–100
Diels-Alder reaction, 103
   of 1,3-butadiene and trifluoroethylene, 104
Diels-Alder synthesis, 102–104
Dienes, conjugated addition to, 102–104
Diethylaminosulfur trifluoride (DAST)
   reaction with β-dicarbonyl compounds, 46
   reaction with ethyl acetoacetate, 5, 45
   replacement of hydroxyl by fluorine, 4, 5, 44–46
Diethyl 2-bromo-3-fluorosuccinate, elimination
   of hydrogen fluoride, 34, 106
Diethyl threo-2-bromo-3-fluorosuccinate, reaction with potassium acetate, 34
Diethyl bromosuccinate, from diethyl 2-bromo-3-
   fluorosuccinate, 106
1,2-Diethylhexafluorocyclopentene, from 1,2-
   dichlorohexafluorocyclopentene, 29, 97
Difluorocarbene
   from chlorodifluoromethane, 75, 83
   insertion into carbon-chlorine bond, 83
   from perfluoropropylene oxide, 103
   reaction with perflorobenzotrichloride, 83
   reaction with phenol and phenoxide, 76
Difluoro compound
   geminal, hydrolysis, 17, 71
   vicinal, from β-dicarbonyl compounds and
   diethylaminosulfur trifluoride, 45, 46
1,1-Difluorocyclohexane, hydrolysis to cyclohexanone, 71
1,2-Difluoro-1,2-diphenylethane, 4
3,3-Difluoro-2-ethoxy-1-phenyl-1-cyclobutene,
   reaction with sulfuric acid, 18, 71
1,1-Difluoroethylene, reaction with trifluoroacetyl hypochlorite, 10, 57

Difluorofumaric acid, catalytic hydrogenation, 6,
   46, 47
Difluoromaleic acid
   hydrogenation, 6
   hydrogenation to fluorosuccinic acid and
   succinic acid, 46, 47
Difluoromethylene group, hydrolysis, 18, 71
Difluoromethylenetris(dimethylamino)phospho-
   rane, from tris(dimethylamino)phosphine
   and dibromodifluoromethane, 27, 94
Difluoromethyl phenyl sulfide, from thiophenoxide and difluorocarbene, 21, 78
3,4-Difluoro-1-nitrobenzene, chlorination, 10, 56
1,3-Difluoropropan-2-ol, cyclization to 3-fluoro-
   1,2-propylene oxide, 75
1,2-Difluorotetrachlorobenzene, reaction with
   sodium methoxide, 24, 88
1,1-Difluorotetrachloroethane, from 1,2-difluo-
   rotetrachloroethane, 109
Difluorobis(thiophenyl)methane, from chlorodifluoromethyl phenyl sulfide, 78
Diglyme, solvent in hydrolysis, 16, 66
1,2-Dimethoxytetrafluorobenzene, from 1,2-
   difluorotetrachlorobenzene, 88
Dimethylamine, reaction with chlorotrifluoroethylene, 23, 85, 86
tetrakis(Dimethylamino)ethylene, from chlorotrifluoroethylene, 85, 86
tris(Dimethylamino)phosphine, 94
   reaction with bromofluoromethanes, 27
tris(Dimethylamino)phosphine oxide, from
   tris(dimethylamino)phosphine, 94
N,N-Dimethylimidazolid-2-one, solvent, 25, 90
Dimethyl malonate, reaction with perfluoroisobutylene, 22, 82
Dimethyltrifluorovinylcarbinol, from trifluoroethylene, 98
Dinitrile, fluorinated, by cleavage of tetrafluoro-
   o-phenylene diamine, 50
2,4-Dinitrochlorobenzene, conversion to 2,4-
   dinitrofluorobenzene, 89, 90
2,4-Dinitrofluorobenzene, from 2,4-dinitro-
   chlorobenzene, 89, 90
2,4-Dinitrohalobenzenes
   low electron density in benzene ring, 89
   reaction with sodium methoxide, 25
N,N'-Diphenylamidine of chlorofluoroacetic acid,
   from chlorotrifluoroethylene and aniline,
   85
1,1-Diphenylethylene
   fluorination, 4
   reaction with aryl iodide difluoride, 43
   reaction with fluorine, 42
Directive effect in aromatic substitution
   of fluorine, 59
   of methyl group, 59
Displacement, nucleophilic, of aromatic
   halogens, 26, 88–91
Displacement of halogen, in halonitro- and halo-
   dinitrobenzenes, reaction rates, 90
Double bond polarization, by fluorine, 51–54
Durene, bromination reaction rate, 56

Elcb (carbanion) mechanism
   in dehydrofluorination of bromofluorocyclo-
   hexane, 107
   in elimination of hydrogen fluoride, 106

Electron density
  of benzene ring, 55, 59, 89
  of carbon adjacent to the trifluoromethyl
    group, 53, 64
  of carbon linked to fluorine, 89
Electron density in fluoroalkenes, 53, 54
Electrophilic addition, to fluoroalkenes, 54
Electrophilic carbon, 52
Electrophilic double bonds, 54
Eliminations, 33, 34, 105–108
E2 mechanism, in dehydrohalogenation, 107
Enamines, reaction with trifluoromethanesulfenyl
    chloride, 23, 83
Enol ethers, reaction with sulfur trioxide, 12, 60
Ethanol, benzoylation with benzoyl trifluoro-
    acetate, 93
Ethanolamine. *See also* 2-Aminoethanol
  nucleophilic addition, 31, 101
2-Ethoxy-1-phenylcyclobut-1-ene-3-one, from
    3,3-difluoro-2-ethoxy-1-phenyl-1-
    cyclobutene, 72
Ethyl acetoacetate
  reaction with 1-chloro-2-fluoroethane, 19, 75
  reaction with diethylaminosulfur trifluoride, 5,
    45
Ethyl 1-acetylcyclopropene-1-carboxylate, from
    1-chloro-2-fluoroethane and ethyl aceto-
    acetate, 75
Ethyl 2-(β-hydroxyethylamino)-3-trifluoro-
    methyl-4,4,4-trifluorobutanoate, 101, 102
Ethyl chlorofluoroacetate
  reaction with Grignard reagents, 28
  reaction with phenylmagnesium bromide, 95,
    96
Ethyl γ,γ,γ-trifluorocrotonate, addition of
    ammonia, 101
Ethylmagnesium bromide, reaction with
    1,2-dichlorohexafluorocyclopentene, 29,
    97
Ethyl 2-oxopentafluoropropane-1-sulfonate, from
    ethyl pentafluoroisopropenyl ether, 60
Ethyl pentafluoroisopropenyl ether, reaction with
    sulfur trioxide, 12, 60
Ethyl 4,4,4-trifluoroacetoacetate, from phos-
    phorus ylides and trifluoroacetonitrile,
    28, 95
Ethyl 4,4,4-trifluoro-2-butenoate, addition of
    ammonia, 101
Ethyl bis(trifluoro)methyl acrylate, reaction with
    2-aminoethanol, 31, 101, 102
Ethyl 4,4,4-trifluoro-3-trifluoromethyl-2-buteno-
    ate, reaction with 2-aminoethanol, 31,
    101, 102
Explosions, of hydrogen fluoride cylinders, 41

Fluorinated aromatic diamines, reaction with
    nitrous acid, 8, 51
Fluorinated compounds, hydrolysis, 17, 65–73
  reduction with hydrogen, 6, 46, 47
  reduction with hydrides, 6, 48
  reduction with isopropyl alcohol, 6, 7, 48, 49
Fluorinated cyclopropanes, reaction with alkalis,
    18, 33
Fluorinated ethylenes, reaction with alkali alkox-
    ides, 20, 77
Fluorination of 1,1-diphenylethylene, with lead
    tetrafluoride, 4, 42

Fluorine
  activation of the benzene ring for electrophilic
    substitution, 55, 59, 60
  addition to unsaturated compounds, 4, 42
  aromatic, hydrolysis, 51
  back-donation of electrons, 51. *See also* Back-
    donation of electrons
  bond dissociation energy, 5, 42
  generation from hydrogen fluoride and mer-
    curic oxide, 43
  from hydrogen fluoride and lead tetraacetate,
    42
  introduction into organic compounds, 4, 5, 42
  migration, 108, 109
  nucleophilic displacement, 24, 51, 90
    from allylic position, 66
    by chloride anion, 56
    by cyano group, 88
    in fluoropyridines, 67
    by methoxy group, 88, 90
    by *p*-methoxyphenoxy group, 90
    by a sulfide ion, 91
  replacement by
    bromine with aluminum bromide, 58
    chlorine with aluminum chloride, 59, 64
Fluorine atoms, in fluorination of 1,1-diphenyl-
    ethylene, 42
Fluoroacetylenes, reaction with sulfur trioxide,
    60, 61
Fluoroalkenes
  reaction with azides, 24, 87
  reaction with trifluoroacetyl hypochlorite, 10, 57
Fluoroaromatic acetylenic ethers, Claisen rear-
    rangement, 37, 111
ω-Fluorocarboxylic acids, hydrolysis, 15, 65
2-Fluorocyclohexanone, reduction with isopro-
    pyl alcohol, 6, 48
Fluoro enol ethers, reaction with sulfur trioxide,
    12, 60
Fluoroform reaction, 69
Fluorofumaric acid, catalytic hydrogenation, 6,
    46, 47
Fluorohaloethanes
  reaction with aluminum chloride, 35
  rearrangements, 108, 109
Fluoroketones, reaction with bromofluoro-
    methanes and phosphines, 27, 94
Fluoromaleic acid, catalytic hydrogenation, 46, 47
2-Fluoro-3-methylbutene-2-oic acid, from
    trifluoroethylene, 98
trans-3-Fluoro-4-methyl-1-phenylsulfonylpenta-
    1,3-diene, from 2,2-difluoro-3,3-dimethyl-
    1-phenylsulfonylmethylcyclopropane,
    105, 106
2-Fluoromethyl-4,5,6,7-tetrafluorobenzofuran, from
    pentafluorophenyl propargyl ether, 111
Fluoronitriles
  hydrolysis, 15, 65
  reaction with phosphorus ylides, 28, 95
2-Fluoro-4-nitrotoluene, from *o*-fluorotoluene, 60
2-Fluoro-5-nitrotoluene, from *o*-fluorotoluene,
    59, 60
2-Fluoro-6-nitrotoluene, from *o*-fluorotoluene, 60
Fluoropentachlorobenzene
  displacement of fluorine, 89
  reaction with potassium *p*-methoxyphenoxide,
    25

Fluoropentanitrobenzene, reaction with hydrogen chloride, 10, 56
3-Fluoro-1,2-propylene oxide, from 1,3-difluoropropan-2-ol, 75
trans-α-Fluorostilbene
  hydrogenation, 6
  hydrogenation to dibenzyl, 46, 47
o-Fluorotoluene
  chlorination, 9, 55
  nitration, 12, 59
2-Fluoro-1,3,3-trichloroindene, from 1-phenylperfluoropropene, 14, 64
Formaldehyde
  reaction with piperidine, 27, 93
  reaction with trifluoroacetone, 27, 93
Four-membered rings, by cycloaddition of fluoroalkenes, 102–104
Free-radical reaction with alcohols, 91
Free-radical mechanism, in addition of hydrogen bromide to fluoroalkenes, 52–54
Friedel-Crafts reaction, 14
Friedel-Crafts synthesis
  of 1-chloro-2-fluoropropane, 62, 63
  of 1-chloro-2-phenylpropane, 63
  of 1-phenyl-3,3,3-trifluoropropane, 64

Geminal difluorides, hydrolysis, 17, 71
Grignard reagent, reaction with chlorofluoroacetates, 28
Grignard synthesis, of chlorofluoromethyl phenyl ketone, 95, 96
Guldberg-Waage's law, 92

Halobenzenes, reaction with nucleophiles, 25, 89, 90
Halogen derivatives, preparation, 8–12, 51–58
Halopyridines
  electrophilic substitutions, 67, 68
  nucleophilic substitutions, 67, 68
Halothane, fluorinated inhalation anesthetic, 108
Heptafluorobutyraldehyde hydrate, hydrolysis, 17, 68
Heptafluoropropane
  from heptafluorobutyraldehyde, 69
  from heptafluoropropyl methyl ketone, 69
Heptafluoropropyl anion, from heptafluorobutyraldehyde, 69
Heptafluoropropyl bromide, from perfluorobutyramide, 110
Heptafluoropropyl group, inductive effect, 69
Heptafluoropropyl methyl ketone, hydrolysis to heptafluoropropane, 69
Heptyl 2*H*-perfluorononanoate, hydrolysis, 71
Hexabromocyclobutene, from hexafluorocyclobutene, 58
Hexachlorocyclobutene, from hexafluorocyclobutene, 58
Hexafluoroacetone azine, reaction with acetylene, 32
Hexafluorobenzene, conversion to hexafluoro Dewar benzene, 110
Hexafluorocyclobutene. *See* Perfluorocyclobutene
  treatment with aluminum bromide, 58
  treatment with aluminum chloride, 58
Hexafluoropropylene oxide. *See* Perfluoropropylene oxide

Hexamethylphosphorous amide. *See also* tris(Dimethylamino)phosphine
  reaction with dibromodifluoromethane, 27, 94
Hofmann degradation, of perfluorobutyramide, 35, 36, 109, 110
Hydrazine
  reaction with 1,2-dichlorohexafluorocyclopentene, 81
  reaction with perfluorocyclobutene, 22, 80, 81
Hydrogen, acidic, in 2*H*-perfluoroisobutane, 95
Hydrogenation
  catalytic of fluorinated compounds, 6, 46, 47
  simultaneous (conjugate) catalytic, 47
Hydrogen bonding, in protonation of fluorine, 65, 72, 17, 108
Hydrogen bonds, 93
Hydrogen bromide. *See also* Hydrogen halides
  addition to chlorotrifluoroethylene, 8, 52
  addition to perfluoropropene, 54
  addition to 1,1,1-trifluoropropene, 53
  free-radical addition to fluoroalkenes, 52–54
Hydrogen chloride
  addition to 3,3,3-trifluoropropene, 9, 53
  reaction with fluoropentanitrobenzene, 10, 56
Hydrogen-chlorine bond, hydrogenolysis, 6, 47
Hydrogen fluoride. *See also* Hydrogen halides
  addition to perfluoropropene, 9, 54, 55
  addition to trifluoropropene, 9, 53
  anhydrous
    reaction with steel cylinders, 3, 41
    storage in steel cylinders, 41
  bond dissociation energy, 42
  vapor pressure, 3
Hydrogen fluoride cylinder safety, pressure in, 3
Hydrogen-fluorine bond, hydrogenolysis, 46
Hydrogen halides
  addition to fluoropropenes, 52–54
  addition to perfluoropropene, 9, 54, 55
Hydrogenolysis
  of chlorine, 48, 49
  of fluorine, 47, 48
  of unsaturated fluoro compounds, 46, 47
Hydrolyses, 15–18, 65–73
Hydrolysis
  of 1-acetyl-2,2-difluro-3-phenylcyclopropane, 72
  of benzyl chloride, 65
  of benzyl fluoride, 65
  of benzyl halides, 15–18
  chlorofluoropyridines, 16
  of 3,5-dichlorotrifluoropyridine, 16, 67
  difluoromethylene group, 18, 71
  in enol ethers, 18, 71, 72
  fluorinated compounds, 15–18, 65–73
  of ω-fluorocarboxylic acid nitriles, 15, 65
  of ω-fluorocarboxylic acids, 15, 65
  ω-fluoronitriles, 65
  geminal difluorides, 17, 71
  heptafluorobutyraldehyde hydrate, 17, 69
  heptyl 2*H*-perfluorononanoate, 17, 71
  1-(*o*-hydroxyphenylpentafluoropropene), 17, 70
  iodopentafluoroethane, 17, 70
  1-(*o*-methoxyphenyl)-perfluoropropene, 17
  pentafluoroethyl iodide, 17, 70
  perchloric acid as catalyst, 65
  perfluoroalkyl iodides, 69, 70

Hydrolysis *(continued)*
  of perfluorocycloalkenes, 16, 66
  perfluoroethyl iodide, 17, 70
  2-phenylhexafluoro-2-propanol, 69
  α,α,α-trifluoroacetophenone, 69
  of 2,5-bis(trifluoromethyl)aniline, 16, 68
  trifluoromethyl group, 68, 69, 73
  of *p*-trifluoromethylphenol, 18, 72, 73
p-Hydroxybenzoic acid, from *p*-trifluoromethyl-
    phenol, 73
Hydroxyl group, activator for hydrolysis of
    trifluoromethyl group, 68
4-Hydroxymethylperfluoropyridine, from
    perfluoropyridine, 91
N-Hydroxymethylpiperidine, from formaldehyde
    and piperidine, 93
4-Hydroxyperfluorobutyryl fluoride cesium salt,
    from cesium and perfluoro-γ-butyro-
    lactone, 77, 78
1-(*o*-Hydroxyphenyl)-2,3,3,3-tetrafluoro-1-pro-
    panone, from 1-(*o*-methoxyphenyl)penta-
    fluoropropene, 70, 71

Inductive effect of halogens, decrease in electron
    density, 89
Inorganic chemists, warmup for, 41, 42
Insertion
  of difluorocarbene, 75, 83
  of sulfur trioxide, 62
Iodine fluoride
  addition to fluoroalkenes, 8, 51, 52
  from iodine pentafluoride and iodine, 51, 52
Iodopentafluoroethane, hydrolysis, 70
Isopropyl alcohol
  reaction with 2-fluorocyclohexanone, 6, 48
  reduction of fluorinated compounds, 48, 49
Isopropylbenzene, from 2-chloropropane and
    benzene, 63, 64

Lactones, intermediates in hydrolysis of
    ω-fluorocarboxylic acids, 65
LDA. *See* Lithium diisopropyl amide (LDA)
Lead tetraacetate
  cleavage of benzene ring, 50
  cleavage of carbon-carbon bonds, 49, 50
Lead tetrafluoride, generation from hydrogen
    fluoride, 42
Lithium aluminum hydride, reduction with, 6, 47,
    48
Lithium diisopropyl amide (LDA), base in reac-
    tions of fluorocyclopropenes, 33

Meisenheimer complex, 67, 88, 89
Mercuric oxide
  oxidation of hydrogen fluoride to fluorine, 43
  in nucleophilic addition of hydrogen fluoride,
    9, 54, 55
Metal fluorides, reaction with perfluoroalkenes,
    9, 54, 55
Methanol
  free-radical addition to perfluoro-1-butene, 86,
    87
  reaction with chlorofluoroalkenes, 23, 84
  reaction with perfluoropyridine, 91
2-Methoxyperfluoro-3,4-dimethyl-3-hexene,
    from perfluoro-3,4-dimethyl-3-hexene,
    77

p-Methoxyphenoxide, reaction with fluoropenta-
    chlorobenzene, 90
p-Methoxyphenoxypentachlorobenzene, from
    fluoropentachlorobenzene, 90
1-(*o*-Methoxyphenyl)perfluoropropene, hydroly-
    sis, 17, 70
Methyl acrylate, addition of ammonia, 101
Methyl 3-chloro-2,3-difluoroacrylate, from
    1-chloro-1,2-difluoroethylene, 99
Methyl 2,3-dichloro-3,3-difluoropropionate, by
    reduction of methyl 2,2,3-trichloro-3,3-
    difluoropropionate, 48
Methyl 2,3-difluoroacrylate, from 1-chloro-1,2-
    difluoroethylene, 99
Methyl 3,3-difluoro-2,2,3-trichloropropionate,
    reduction with isopropyl alcohol, 6, 48,
    49
Methyl 2-methoxycarbonyl-5,5,5-trifluoro-4-tri-
    fluoromethylpenta-2,3-dienoate, from
    perfluoroisobutylene, 82
Methyl 4-oxo-2-phenylpentanoate, from 1-acetyl-
    2,2-difluoro-3-phenylcyclopropane, 72
2-Methyl-3,4,4-trifluoro-3-buten-2-ol, from tri-
    fluoroethylene, 98
Michael addition, of 2*H*-perfluoroisobutane and
    acrolein, 95
Migration of fluorine, 108, 109
1-(*N*-Morpholinyl)-5-phenylaminocarbonyl-
    cyclopentene, from 2-phenylamino-
    carbonylcyclopentanone, 83, 84

Nitration, 12, 59, 60
  of *o*-fluorotoluene, 12, 59, 60
Nitriles of ω-fluorocarboxylic acids, hydrolysis, 15
3-Nitro-1,4-bis(trifluoromethyl)benzene, from
    4-chloro-3-nitrobenzotrifluoride, 100
Nitrohalobenzenes, nucleophilic displacement of
    halogens, 25
Nucleophiles, reaction with halobenzenes, 25, 69
Nucleophilic addition. *See also* Additions, elec-
    trophilic; Additions, free-radical; Addi-
    tions, nucleophilic
  of ethanolamine, 101
  of hydrogen fluoride, 54, 55
  of perfluorocyclobutenes, 80
Nucleophilic displacement
  of aromatic fluorine, 51, 88
  of fluorine by chloride anion, 56

Octafluoroisobutylene. *See* Perfluoroisobutylene
Olefins. *See* Alkenes
Organometallic complexes, of 1,1-dichloro-2,2,2-
    trifluoroethylzinc chloride, 98–100
Organometallic syntheses, 28–31, 95–100
Oxidation, 7, 8, 49–51
2-Oxopentafluoropropanesulfonic acid, by trans-
    esterification, 60
2-Oxo-1,1,3,3,3-pentafluoropropyl fluorosulfate
  from perfluoropropylene oxide, 62
  by rearrangement of perfluoro-1,2-propanediol
    sulfate, 62

2*H*-Pentafluorocyclopentane-1,3-dione, from
    perfluorocyclopentene, 66, 67
Pentafluoroethane, by hydrolysis of pentafluoro-
    ethyl iodide, 70

Pentafluoroethyl iodide, hydrolysis, 17, 69, 70
Pentafluorophenol
    reaction with chlorodifluoromethane, 20, 75,
        76
    reaction with difluorocarbene, 75, 76
    reaction with tert-butyl hypobromite, 11, 57
Pentafluorophenoxycarbene, from pentafluoro-
        phenol and carbene, 76
bis(Pentafluorophenoxy)fluoromethane, from
        pentafluorophenol and carbene, 76
Pentafluorophenyl propargyl ether, rearrange-
        ment, 37, 111
1H-Pentafluoropropene, reaction with azides, 87
Pentafluoropyridine. See Perfluoropyridine
bis(Pentaflurophenoxy)fluoromethane, 76
Perchloric acid, catalyst in hydrolysis, 65
Perfluoroalkenes
    addition of alcohols, 86, 87
    addition of free-radicals, 86, 87
    polarization of double bond, 54
    reaction with alcohols, 24
    reaction with metal fluorides, 9, 54, 55
Perfluoroalkyl group, inductive effect, 69, 70
Perfluoroalkyl iodides, hydrolysis, 69, 70
Perfluorobenzotrichloride, reaction with chloro-
        difluoromethane, 22, 83
Perfluoro-1,3-butadiene, dimerization, 33, 105
Perfluoro-1-butene, reaction with methanol, 24,
        86, 87
Perfluoro-tert-butylacetylene, reaction with sulfur
        trioxide, 13, 60
Perfluoro-tert-butylketene-α-sulfonyl fluoride,
        from fluoro-tert-butylacetylene and SO₃,
        60, 61
Perfluoro-2-butyltetrahydrofuran, reaction with
        aluminum chloride, 11
Perfluoro-2-butyl-2,5,5-trichlorofuran, from
        perfluoro-2-butyltetrahydrofuran by
        aluminum chloride, 58
Perfluorobutyramide, Hofmann degradation, 35,
        36, 109, 110
Perfluoro-γ-butyrolactone, reaction with per-
        fluoropropylene oxide, 21, 77, 78
Perfluorocycloalkenes
    hydrolysis, 16, 66, 67
    reaction with aluminum halides, 11, 58, 59
Perfluorocyclobutene
    reaction with hydrazine, 22, 80, 81
    replacement of fluorine by halogens, 11, 58
Perfluorocyclopentenes, hydrolysis, 66
Perfluorocyclopenten-1-ol, from perfluorocyclo-
        pentene, 66, 67
Perfluorocyclopent-2-en-1-one, from perfluoro-
        cyclopentene, 66
Perfluorodecalin, reaction with sodium phen-
        oxide, 21
Perfluoro-2,5-dihydrotetramethylfuran, from per-
        fluoro-3,4-dimethyl-3-hexene, 77
Perfluoro-3,4-dimethyl-3-hexene, reaction with
        methanol and pyridine, 20, 77
Perfluoroethyl iodide, hydrolysis, 70
Perfluorohexamethylbenzene, rearrangement, 110
Perfluoroindane, from 2-bromoperfluoro-
        naphthalene, 111
Perfluoroisobutylene
    reaction with dimethyl malonate, 22, 82
    reaction with sulfur trioxide, 13, 61

Perfluoronaphthalene, reaction with
    sodium alkoxides or aryloxides, 25
    sodium salt of 2-hydroxytetralin, 90, 91
    sodium salt of m-cresol, 90, 91
    sodium salt of 2-naphthol, 90, 91
Perfluoro-o-phenylenediamine, reduction with
        lead tetraacetate, 7
1H,1H,3H-Perfluoropentanol, from perfluoro-1-
        butene and methanol, 87
5H-Perfluoro-4-pentenoic acid, from perfluoro-
        cyclopentene, 66
Perfluoropropene
    addition of hydrogen halides, 9, 53, 54
    addition to iodine fluoride, 52
    nucleophilic addition of hydrogen fluoride, 54,
        55
    polarization of double bond, 53, 54
Perfluoropropylene oxide
    ejection of difluorocarbene, 103
    reaction with 1,2-dichlorodifluoroethylene, 32
    reaction with perfluoro-gamma-butyrolactone,
        21, 77, 78
    reaction with sulfur trioxide, 13, 62
Perfluoropropylisocyanate, from perfluorobutyr-
        amide, 110
Perfluoropyridine, reaction with methanol, 26, 91
Perfluorotetrahydropyran, reaction with alumi-
        num chloride, 12, 58, 59
Perfluorotoluene, reaction with sulfur trioxide,
        13, 62
Perfluorotricyclooctane, by dimerization of per-
        fluoro-1,3-butadiene, 105
Perfluorovinyl. See Trifluorovinyl
Perfluorovinylsulfur pentafluoride
    reaction with 1,3-butadiene, 32, 103
Phenol, benzoylation with benzoyl trifluoro-
        acetate, 93
2-Phenylaminocarbonylcyclopentanone enamine
        with morpholine, 23, 83, 84
2-Phenylhexafluoro-2-propanol, hydrolysis, 69
Phenyl iodide difluoride, fluorination agent, 4
Phenylmagnesium bromide
    reaction with chlorofluoroalkenes, 29, 96
    reaction with ethyl chlorofluoroacetate, 28, 95,
        96
2-Phenyl-1,1,3,3,3-pentafluoropropene, by Wittig
        reaction, 94
3-Phenylpentafluoropropene, from 3-chloro-
        pentafluoropropene, 96
1-Phenylperfluoropropene, reaction with alumi-
        num chloride, 14, 64
octakis(Phenylthio)naphthalene, from perfluoro-
        decalin, 21
1,4,5,8-tetrakis(Phenylthio)perfluorooctalin, from
        perfluorodecalin and sodium thiophen-
        oxide, 79
2-Phenyl-3H-perfluorohex-2-ene, from α,α,α-tri-
        fluoroacetophenone and tributylphos-
        phine, 94
Phenyl trifluoromethyl sulfide, from thiophen-
        oxide and bromotrifluoromethane, 79
1-Phenyl-3,3,3-trifluoropropane, from 3,3,3-tri-
        fluoropropene, 64
Phosphines, reaction with carbonyl compounds, 94
Phosphorus ylides
    reaction with trifluoroacetonitrile, 28, 95
    in Wittig reaction, 94

Piperidine, reaction with formaldehyde, 27, 93
Polarity reversal, by trifluoroethyl group, 101, 102
Polarization of double bond
    by fluorine, 51
    in fluoroalkenes, 57
Polyfluoroalkyl carbanion, from polyfluoroalkyl
    halides, 78
Polyfluorohalomethane
    conversion to carbenes, 78
    reaction with alkali thiophenoxides, 21
Poly(trifluoromethyl)benzenes, reaction with
    ammonia, 22
Potassium acetate, elimination of hydrogen
    fluoride, 106
Potassium tert-butoxide, reaction with bromo-
    fluorocyclohexanes, 34
Potassium fluoride, in nucleophilic addition of
    hydrogen fluoride, 54, 55
Preparation of halogen derivatives, 8–12, 51–58
Pressure, build-up in hydrogen fluoride cylinder,
    41

Reaction rate
    of bromination of benzene homologs, 56
    in displacement of halogen from halonitro-
        benzenes, 89
Rearrangement, 35–37, 108–111
    of 1-bromo-2-chloro-1,1,2-trifluoroethane, 108
    of 2-bromoperfluoronaphthalene, 111
    of 1-chloro-1,2-dibromotrifluoroethane, 109
    of 2-chloro-1-methyl-1-phenylethylene oxide,
        96
    of 3-chloroperfluoropropene, 109
    ortho-Claisen, 111
    of 1,2-difluorotetrachloroethane, 109
    in fluorination of 1,1-difluoroethylene, 42
    of 1,1-bis(p-chlorophenyl)-2,2,2-trichloro-
        ethane (DDT), 43
    of pentafluorophenyl propargyl ether, 111
    of perfluorohexamethylbenzene, 110
    of perfluoro-1,2-propanediol sulfate, 62
    of 1-phenylperfluoropropene, 64
    in reaction of diethylaminosulfur trifluoride
        with saccharides, 44
    in reaction of trifluoronitrosomethane with
        ammonia, 109
    in replacement of hydroxyl by fluorine, 4
    of 1,1,2-trichlorotrifluoroethane, 109
    of trifluoroisopropyl cation, 63
Reduction, 6, 7, 46–49
    with complex hydrides, 6, 7, 46–49

Saccharides
    reaction with diethylaminosulfur trifluoride,
        44, 45
    replacement of hydroxyls by fluorine, 4, 5
S$_{AR}$1 mechanism, 79
Silver fluoride, addition to perfluoropropene, 55
N-Silver perfluorobutyramide, from perfluoro-
    butyramide, 110
Simultaneous (conjugate) catalytic hydrogena-
    tion, 47
S$_N$Ar mechanism, 90
S$_N$i mechanism
    in cyclization of ethyl 5-fluoropentan-2-one-3-
        carboxylate, 75
    in hydrolysis of ω-fluorocarboxylic acids, 65

S$_N$i mechanism (continued)
    in treatment of saccharides with diethylamino-
        sulfur trifluoride, 44, 45
S$_N$2 mechanism, in hydrolysis of ω-fluoronitriles,
    65
S$_N$2 nucleophilic displacement, 75
S$_N$2 reaction
    opening of fluorocyclopropane ring, 105
    in treatment of saccharides with diethylamino-
        sulfur trifluoride, 44, 45
Sodamide, reaction with cis-1-bromo-2-fluoro-
    cyclohexane, 34
Sodium azide, reaction with fluoroalkenes, 24, 87
Sodium methoxide, reaction with bromofluoro-
    cyclohexanes, 34
Sodium sulfide, reaction with 2-bromo-4,5-
    difluoronitrobenzene, 26, 91
Sodium thiophenoxide, reaction with perfluoro-
    decalin, 79
cis-Stilbene, fluorination, 4
Sulfur trioxide
    addition to fluoroacetylenes, 60
    electrophilic substitution, 55, 56, 59, 62
    reaction with fluorinated compounds, 12, 13
    reaction with fluoro enol ethers, 60
    reaction with perfluoroisobutylene, 61
    reaction with perfluoropropylene oxide, 62
    reaction with perfluorotoluene, 13, 62
Sulfur trioxide reactions, 60–62

Tetraethylammonium fluoride, in nucleophilic
    addition of hydrogen fluoride, 54, 55
Tetrafluoroethylene
    bond strain, 102
    F–C–F bond angle, 102
    sp^3 bonds, 102
Tetrafluoro-2,4-hexanedioic dinitrile, from
    1,2-diaminotetrafluorobenzene, 50
Tetrafluoro-o-phenylenediamine, reduction with
    lead tetraacetate, 7, 49, 50
2,3,6,7-Tetrafluoro-1,4,5,8-tetrakis(phenylthio)-
    naphthalene, from perfluorodecalin and
    sodium thiophenoxide, 79
2,4,5,6-Tetrafluoropyridine, from 3-chlorotetra-
    fluoropyridine by catalytic hydrogena-
    tion, 47, 48
1,2,4,5-Tetramethylbenzene, bromination reac-
    tion rate, 56
Thiophenoxide
    reaction with dichlorodifluoromethane, 78
    reaction with perfluorodecalin, 21
Tributylphosphine, reaction with α,α,α-trifluoro-
    acetophenone, 27, 94
5,5,5-Trichlorohexafluoropentanoyl chloride,
    from perfluorotetrahydropyran by alumi-
    num chloride, 58, 59
1,1,1-Trichlorotrifluoroethane
    formation of 1,2-dichloro-2,2,2-trifluoroethyl-
        zinc chloride, 98–100
    reaction with zinc and aldehydes, 30
    from 1,1,2-trichlorotrifluoroethane, 109
2,4,6-Tricyanobenzotrifluoride, from 1,2,3,5-
    tetrakis(trifluoromethyl)benzene, 81, 82
Triethylammonium azide, reaction with fluoro-
    alkenes, 24, 87
Triethyl chlorofluoroorthoacetate, from chloro-
    trifluoroethylene, 77

Trifluoroacetic acid anhydrides, acylation with, 26
Trifluoroacetic anhydrides, mixed, acylating
    agents, 92, 93
1,1,1-Trifluoroacetone, reaction with piperidine,
    27, 93
1,1,1-Trifluoroacetone hydrate, reaction with
    N-hydroxymethylpiperidine, 93
Trifluoroacetonitrile
    reaction with phosphorus ylides, 95
    reaction with tris(tert-butyl)azete, 33, 104, 105
α,α,α-Trifluoroacetophenone
    hydrolysis, 69
    reaction with difluoromethylenetris(dimethyl-
        amino)phosphorane, 94
    reaction with tributylphosphine, 27, 94
Trifluoroacetyl fluoride, from perfluoropropylene
    oxide, 103
Trifluoroacetyl hypochlorite
    addition to fluoroalkenes, 57
    dissociation, 57
    polarity of double bond, 57
    reaction with 1,1-difluoroethylene, 57
2,2,2-Trifluoroethanol, benzoylation with
    benzoyl trifluoroacetate, 93
Trifluoroethylene. See also Fluoroalkenes
    cycloaddition, 103, 104
    reaction with 1,3-butadiene, 32, 103, 104
    reaction with butyllithium, 30, 98
Trifluoromethanesulfenyl chloride, reaction with
    enamines, 23, 83, 84
2,5-bis(Trifluoromethyl)aniline, hydrolysis, 16, 68
1,2,3,5-tetrakis(Trifluoromethyl)benzene
    ammonolysis, 81, 82
    conversion to 2,4,6-tricyanobenzotrifluoride,
        81, 82
    reaction with ammonia, 22
hexakis(Trifluoromethyl)benzene
    isomerization by irradiation, 36
    rearrangement to perfluorobenzvalene, 110
    rearrangement to perfluorohexamethyl Dewar
        benzene, 110
    rearrangement to perfluoroprismane, 110
Trifluoromethylcopper
    generation and reactions, 31
    reaction with 4-chloro-3-nitrobenzotrifluoride,
        100
2,2-bis(Trifluoromethyl)-1-fluoroethenyl fluoro-
    pyrosulfate, from perfluoroisobutylene, 61

2,2-bis(Trifluoromethyl)-1-fluoroethenyl fluoro-
    sulfate, from perfluoroisobutylene and
    sulfur trioxide, 61
Trifluoromethyl group
    ammonolysis, 81, 82
    conversion to nitrile group, 81, 82
    drainage of electrons from double bond, 53
    hydrolysis, 16, 18, 68, 73
    reaction with ammonia, 22
bis(Trifluoromethyl)hydroxylamine, from tri-
    fluoronitrosomethane and ammonia, 109
tris(Trifluoromethyl)methane, reaction with
    acrylonitrile, 31
p-Trifluoromethylphenol, hydrolysis, 18, 73
4-Trifluoromethyl-5,5,5-trifluorovaleronitrile,
    from 2H-perfluoroisobutane and acrolein,
    95
Trifluoronitrosomethane, reaction with ammo-
    nia, 35, 109
3,3,3-Trifluoropropane, reaction of benzene with,
    14
1,1,1-Trifluoropropene
    addition of hydrogen halides, 53
    polarization of double bond, 53
3,3,3-Trifluoropropene, addition of hydrogen
    bromide and chloride, 9, 53
2,3,3-Trifluoro-1-vinylcyclobutane, 103
2,3,3-Trifluoro-1-vinyl-2-cyclobutylsufur penta-
    fluoride, from trifluorovinylsulfur penta-
    fluoride, 103
Trifluorovinyllithium, from trifluoroethylene and
    butyllithium, 98
Trifuoroacetyl hypochlorite, reaction with
    1,1-difluoroethylene, 10, 57
Triple bond, addition of sulfur trioxide, 61

α,β-Unsaturated esters, polarity reversal, 101,
    102
Unsaturated fluorinated compounds, reduction of,
    6, 46–48

Vapor pressure of hydrogen fluoride, 3

Wheland complex, 57
Wittig reaction, 94

Ylides. See Phosphorus ylides